Biological Nomenclature

Crane Russak

First published 1973
by Edward Arnold (Publishers) Limited,
25 Hill Street,
London, W1X 8LL
in conjunction with
The Systematics Association

Second edition 1977
Second edition first published in USA by
Crane, Russak & Company, Inc.
347 Madison Avenue, New York, N.Y. 10017

ISBN 0 8448 1264 1

Library of Congress Catalog Card no.: 77 90821

To
Arthur Allman Bullock
kind mentor in matters nomenclatural
and contributor to a better Botanical Code

Printed in Great Britain

Foreword

Although scientific names are very widely used by the biological community and many others in non-biological fields, few people, apart from professional taxonomists, have more than a passing understanding of the principles governing biological nomenclature. The official Codes of Nomenclature are forbidding documents and daunting to use without special training or guidance. It is largely for these reasons that various approaches have been made to the Systematics Association in recent years to sponsor the production of a relatively simple guide to the principles of biological nomenclature and the workings of the various Codes.

The Council of the Systematics Association willingly agreed to sponsor such a work and was fortunate in being able to persuade Mr. C. Jeffrey, of the Herbarium, Royal Botanic Gardens, Kew, to prepare a suitable text. Mr. Jeffrey's book is a lucid and highly readable account of the subject and he is to be warmly congratulated on providing an eminently practical guide to a highly complex field. In addition, his first chapter, which outlines the general context of systematics, is one of the clearest expositions available, and the glossary/index will be widely consulted.

In sponsoring this work the Systematics Association is confident that Mr. Jeffrey's book will go a long way towards clarifying one of the most intimidating areas of biological systematics and will be a major contribution to communication and understanding between biologists in many disciplines.

V. H. Heywood
formerly President, Systematics Association
1973

Preface

The purpose of this handbook is to provide a practical guide to the use of the nomenclatural parts of taxonomic literature, to promote understanding of the problems, principles and practice of biological nomenclature and to act as an introduction to the Codes of Nomenclature themselves. It is not intended to be used as a substitute for the Codes and the interpretation of any provision of any Code is in no way to be taken as authoritative or definitive. Every effort has been made, however, to ensure factual accuracy and to present what were at the time of writing nomenclaturally orthodox views. Even so, since the Codes are subject to modification, it is inevitable that a few of the details will in time become obsolete. This is especially likely in the fields of virology and bacteriology; the first has as yet no definitive Code of Nomenclature, and a new edition of the Bacteriological Code will be published within the next two years. The Zoological and especially the Botanical Codes are unlikely to be subjected to much alteration, and general principles are in all cases unlikely to be changed.

To the following, who kindly read through the draft text, I am grateful for corrections and helpful suggestions: G. C. Ainsworth, R. K. Brummitt, J. S. L. Gilmour, J. Lewis, S. P. Lapage, K. McKenzie, R. J. Pankhurst, P. H. A. Sneath, B. T. Styles, P. Whitehead and P. F. Yeo. To J. G. Sheals and J. D. Turton I am indebted for helpful advice and comments. John Lewis I wish also to thank for general help and for liaison with the Systematics Association and the Institute of Biology. Responsibility for all errors and omissions remains, however, entirely mine. To A. J. Boyce I am indebted for duplication and distribution of the draft text. Finally, my best thanks are given to my colleague, Mrs. J. S. Page, for typing my manuscript, for reading the proofs and for eliminating much that was obscure, verbose, repetitive and tedious.

C. J.

Kew
1973

Preface to the Second Edition

Since the first edition was published in 1973, the following developments in nomenclature have occurred. A summary of modifications to the Zoological Code made since the 16th International Congress of Zoology in 1963 has been published in *Bull. Zool. Nomencl.*, **31**: 77–101 (1974) and **32**: 65 (1975); an International Botanical Congress in Leningrad in June 1975 has made some small amendments to the Botanical Code; a radically revised Bacteriological Code has been published (Dec. 1975); revised rules of viral nomenclature have been accepted by the International Committee on the Taxonomy of Viruses and published in *J. Gen. Virol.*, **31**: 463–470 (1976); and guidelines for the naming of plant varieties, approved by the Council of the Union for the Protection of New Varieties of Plants (UPOV) in Oct. 1973, have been published in *Plant Varieties and Seeds Gazette*, **109**: 1–3 (1974). Amendments to the text have been made where necessary to accommodate these changes. Opportunity has also been taken to amend certain passages and entries to the glossary that reviewers and others have shown to be inaccurate or unclear.

In addition to the persons mentioned in the original preface, I am also indebted to the following for helpful discussion and advice: F. Fenner, J. L. Melnick, J. W. B. Nye, R. V. Melville, D. Heppell, E. G. Voss, and my colleague F. G. Davies.

C. J.

Kew
1976

Contents

*NOTE

Certain additional information, supplementing but not essential to the main thesis, is given in a section 'Notes to the Text' (p. 51). The small 'superior' figures ([1], [2], . . .) given in the main text refer to the corresponding numbers of these notes.

1

The Systematic Background

1.1 Systematics

The earth is unique among the planets we know in supporting a vast array of *living organisms* of the most diverse kinds. Together with the non-living components of their environment with which they inter-react they have produced and maintain the planetary *ecosphere*. Man himself is a part of the ecosphere and his survival depends upon its continued operation. We are more directly dependent upon some living organisms—e.g. the major food crops and the species of commercial fisheries—than others, but all are important as components of the ecosphere, and have become the objects of study of the field of human endeavour known as *biology*.

One of the first tasks of biology was to make meaningful generalizations about living organisms so that useful knowledge could be passed on from person to person and human behaviour regulated in its light. Early in human history it was found useful to know in advance, for example, what animals were dangerous, what were good to hunt for food, what plants were poisonous and so on. It was soon noticed that living organisms possessed certain consistent features by which they could be reliably identified and sorted into constantly and recognizably distinct groups. Properties like dangerousness, edibility and poisonousness could thus be reliably inferred and the possibly unpleasant consequences of a trial and error approach avoided.

The refinement of this process of recognition and grouping into the scientific study of the diversity of living organisms has given rise to the branch of biology known as *systematics*.[1] The task of systematics is to produce systems of classification which best express the various degrees of overall similarity between living organisms. Such systems are used in biology for the storage, retrieval and communication of information and for the making of reliable predictions and generalizations. They are based on as broad as possible study of the variation of living organisms and aim to establish groups, the members of which possess the largest number of common features and exhibit therefore the greatest overall similarity.

The possibility of constructing such systems, of course, depends upon the occurrence of different features associated in definite combinations in different living organisms. If features all varied independently of one another, then each feature considered would produce a different way of grouping organisms and no one grouping based on greatest overall similarity would be possible. However, this is not so and it is possible to construct systematic groupings that are based on multiple correlations of common features and which reflect greatest overall similarity. This is in general a result of the fact that all living organisms are related to one another to a greater or lesser degree by way of evolutionary descent, and it is this evolutionary relationship that makes possible the establishment of meaningful systematic groupings.

1.2 Classification and nomenclature

Two major fields of systematics are *classification* and *nomenclature*. Classification is the process of establishing and defining systematic groups.[2] The systematic groups so produced are known as *taxa* (singular, *taxon*). Nomenclature is the allocation of names to the taxa so produced. In carrying out their researches, systematists first complete their classificatory work. Only when they are sure they have achieved, on the basis of the information available, the best possible systematic arrangement of the organisms they have studied, do they begin to ascertain the correct names for the taxa they have established. In other words, classification precedes naming, and nomenclature is to this extent independent of classification. Nevertheless, it is necessary first to consider certain aspects of the classification of living organisms which are essential to the understanding of the way in which they are named.

1.3 The taxonomic hierarchy

If we study the living organisms existing in a particular place at a particular time, we find that they occur as series of similar individuals showing certain common features. Such series of recognizably similar individuals, recognizably distinct from other such series, are in general what the systematists call *species*. In sexually reproducing organisms it is also found, in general, that individuals of a species are inter-fertile with one another but reproductively isolated from individuals of other species. When species are compared with one another, it is found convenient to group together those with most features in common into larger, more inclusive taxa which are called *genera*. Genera are in their turn grouped likewise into yet more inclusive taxa called *families*, and so on. Such an arrangement of taxa into an ascending series of ever-increasing inclusiveness forms what is known as an *hierarchical*[3] *system* of classification. In an hierarchical system we start at the bottom with individuals and end up at the top with one all-embracing

Table 1 The categories of the taxonomic hierarchy

This shows the categories of the taxonomic hierarchy usually employed in Botany, Bacteriology and Zoology. They are given their recognized Latin names (often anglicized as in the right-hand column) and are arranged in the relative order in which they must be employed. The most important categories are given in CAPITALS, those seldom used are enclosed in parentheses (Divisio). The categories *Divisio* and *Subdivisio* of the Botanical and Bacteriological Codes correspond to, and are used in place of, the categories *Phylum* and *Subphylum* respectively of zoological usage

Botanical	Bacteriological	Zoological	English Equivalent
REGNUM	REGNUM	REGNUM	Kingdom
		Subregnum	Subkingdom
		(Superphylum)	Superphylum
DIVISIO	(Divisio)	PHYLUM	Division/Phylum
Subdivisio	(Subdivisio)	Subphylum	Subdivision/ Subphylum
		Superclassis	Superclass
CLASSIS	CLASSIS	CLASSIS	Class
Subclassis	(Subclassis)	Subclassis	Subclass
		Infraclassis	Infraclass
(Superordo)		Superordo	Superorder
ORDO	ORDO	ORDO	Order
(Subordo)	(Subordo)	Subordo	Suborder
		Infraordo	Infraorder
		Superfamilia	Superfamily
FAMILIA	FAMILIA	FAMILIA	Family
Subfamilia	(Subfamilia)	Subfamilia	Subfamily
		(Supertribus)	Supertribe
Tribus	Tribus	Tribus	Tribe
Subtribus	(Subtribus)	Subtribus	Subtribe
GENUS	GENUS	GENUS	Genus
Subgenus	(Subgenus)	Subgenus	Subgenus
Sectio			Section
Subsectio			Subsection
Series			Series
Subseries			Subseries
SPECIES	SPECIES	SPECIES	Species
Subspecies	(Subspecies) (=Varietas)	Subspecies	Subspecies
Varietas			Variety
(Subvarietas)			Subvariety
Forma			Form
(Subforma)			Subform

taxon. In between we have various taxa of organisms at different levels of the hierarchy, each of which is subordinate to one and only one immediately higher taxon and each of which (except the lowest) includes one or more subordinate lower taxa.

The arrangement of taxa into an hierarchical system had its origin in the logical theory of classification. It functions primarily as an aid to memory, but it also has a biological basis, in so far as the various levels in the hierarchy can be said to reflect different degrees of evolutionary divergence. The number of levels in the hierarchy, needed conveniently to accommodate the variation of the living world, has none the less been decided quite arbitrarily as a result of practical experience over the past two hundred years. Those generally employed are shown in Table 1. Additional levels may be employed if required. The levels are given conventional names and arranged in a conventional order which must be strictly adhered to. The framework thus formed is known as the *taxonomic hierarchy*. The different levels are known as *taxonomic ranks*. All such taxa as stand at any given level (or rank) in the hierarchy are said to belong to the same *taxonomic category*.

The taxonomic hierarchy can be envisaged as a series of containers, with adjacent walls and bases, placed one inside another, and differing only in height. The containers themselves then represent the taxonomic categories. The levels of the roofs of the containers represent the taxonomic ranks. The contents of the containers—the groups of organisms we place in them—represent the taxa. This analogy also makes it easier to appreciate that taxonomic categories and ranks are purely abstract concepts. It is the taxa—groups consisting ultimately of individual living organisms—that alone have any concrete basis.[4] Thus all the primroses form a taxon which is considered to be of specific rank and is therefore assigned to the category species. This taxon is the species known as *Primula vulgaris*. Similarly, *Primula* is a genus, a taxon of generic rank, assigned to the category genus; and *Primulaceae* is a family, a taxon of family rank, assigned to the category family.[5]

2
Names and Codes

2.1 The purpose of names

A name is merely a conventional symbol or cipher, which serves as a means of reference and avoids the need for continuous use of a cumbersome descriptive phrase. The purpose of names is to act as vehicles of communication. Like the ciphers of any code, names can effectively fulfil this function only if they are understood by, and have the same meanings for, all who use the code. Names, however communicated, should immediately and unequivocally call to mind the concepts intended by the transmitter of the names. This is a fundamental principle of nomenclature and it is the most important criterion by which the efficiency of any system of nomenclature can be judged. It implies that names must be unambiguous and universal.

2.2 Codes of nomenclature

Common names of living organisms in vernacular languages are, in general,[6] so far from meeting these conditions that they are quite unsatisfactory for use in biological nomenclature. Quite apart from the multiplicity of languages, many using different alphabets, even within a single language the same name is often used in different senses to denote different kinds of organisms, or the same kind of organism is known by more than one name. Biological nomenclature tries to avoid such defects, and for this reason sets of rules called *Codes of Nomenclature* have been drawn up. The formation and use of the scientific names of organisms classified as animals are governed by the International Code of Zoological Nomenclature (ICZN); of those classified as plants (including fungi) by the International Code of Botanical Nomenclature (ICBN); and of those classified as bacteria (including actinomycetes) by the International Code of Nomenclature of Bacteria (ICNB).

The three codes differ in approach and format but the operative core of each consists of a series of numbered *rules* or *articles*, some of which are supplemented by *recommendations*. The provisions of rules are mandatory and must be followed whenever names are given or employed. Recommendations deal with subsidiary points and indicate the best procedure to be followed. Names contrary to a recommendation may not be rejected on that count, but they are not examples to be followed. The rules of the Codes do not, of course, have any legal status in national or international law. Their enforcement depends solely on the voluntary agreement of systematists to observe their provisions. The only sanctions that can be employed against those who do not are disapproval by their colleagues and disregard of their work. Nevertheless, non-observance of the provisions of the Codes can lead only to instability of nomenclature. All systematists should therefore understand the provisions of the appropriate Code and follow them even if, personally, they disagree with some of them. This does not preclude the proposal of modifications or exceptions to the rules through the appropriate established procedure.

2.3 Modification of the codes

The Botanical Code may be modified only by a decision of a plenary session of an International Botanical Congress on a resolution made by the Nomenclature Section of the Congress. Permanent Nomenclature Committees are elected by a Congress and are established under the auspices of the International Association for Plant Taxonomy to deal with various nomenclatural matters referred to them. Of these, the Editorial Committee is charged with the preparation and publication of the Code in conformity with the decisions adopted by a Congress. Proposals for modification of the Code must be submitted to the Nomenclature Section of a Congress and are voted on in accordance with a set procedure.

The Bacteriological Code may be modified only by action of the International Committee on Systematic Bacteriology[7] on proposals made to it by its Judicial Commission. The Judicial Commission is elected from the membership of the International Committee and is responsible through an Editorial Board for the editing and production of the Code. Proposals for modification of the Code should be submitted to the Editorial Secretary of the International Committee, who is also Secretary of the Judicial Commission.

The Zoological Code may be modified only by an International Congress of Zoology, or recognized equivalent (currently the Division of Zoology of the International Union of Biological Sciences, at a General Assembly of the International Union of Biological Sciences) to which an International Congress of Zoology has delegated such powers, acting on a recommendation from the International Commission on Zoological Nomenclature presented through and approved by the Section on Nomenclature of the Congress or recognized equivalent. The Code is prepared on behalf of the International Commission by an Editorial Committee appointed by the Congress or recognized equivalent. Proposals for the modification of the

Code should be submitted to the Secretary of the International Commission at least one year in advance of the next International Congress or recognized equivalent. The Zoological Code may also be provisionally modified between Congresses (or their equivalents) by means of *declarations* of the International Commission on Zoological Nomenclature (see §5.16, p. 25).

3
Scientific Names

3.1 Alphabet and language

The Codes of Nomenclature differ in detail but certain basic features are common to all three. To be universal, scientific names must be written in the same alphabet and the same language. The Codes of Nomenclature require that all scientific names be *Latin* in form, written in the Latin alphabet and subject to the rules of Latin grammar.[8] The scientific names of living organisms are therefore Latin or are treated as Latin, even if, as is often the case, they are derived from other languages. The Codes also lay down a number of conventions which must be observed in the formation and use of scientific names so that uniformity is as far as possible ensured.

3.2 Names of taxa above the rank of genus

The names of taxa above the rank of genus consist of one term only and are therefore called *uninomial, uninominal* or *unitary.* They are plural nouns (or adjectives used as nouns) and are written with a capital initial letter. So that the rank of a taxon may be apparent from its name, in many cases the Codes stipulate a standardized ending for the names of all taxa of a given taxonomic rank. For example, under the Botanical Code, the names of plant families must end in *-aceae,*[9] while under the Zoological Code, the names of animal families must end in *-idae.* Such standardized endings as are required by the Codes are listed in Table 2. It should be noted that the names of taxa above the rank of *Superfamilia* are not governed by the Zoological Code (see Table 1).

3.3 Names of genera

The names of genera are also uninomial. They are singular nouns written with a capital initial letter, e.g. *Primula, Felis, Agaricus, Bacillus.*

3.4 Names of taxa intermediate in rank between genus and species

Under the Bacteriological and Zoological Codes, only one category of such taxa—the subgenus—is recognized. The names of subgenera under these

Table 2 Standardized Endings for the Names of Taxa

Endings enclosed in parentheses are only recommended and are not mandatory under the respective Code. The endings *-phyta* and *-phytina* are used for the names of taxa of green (non-fungal) plants; *-phyceae* and *-phycideae* for the names of taxa of algae; *-opsida* and *-idae* for the names of taxa of higher green plants; *-mycota*, *-mycotina*, *-mycetes* and *-mycetidae* for the names of taxa of fungi. The endings *-ales* and *-ineae* are mandatory under the Botanical Code for the names of orders and suborders respectively only if they are names based on the name of an included family. The ending *-acea*, although not reccmmended by the Zoological Code, is also frequently used for the names of superfamilies

Category	Botanical	Bacteriological	Zoological
Divisio	(-phyta/-mycota)		
Subdivisio	(-phytina/-mycotina)		
Classis	(-phyceae/-mycetes/-opsida)		
Subclassis	(-phycidae/-mycetidae/-idae)		
Ordo	-ales	-ales	
Subordo	-ineae	-ineae	
Superfamilia			(-oidea)
Familia	-aceae	-aceae	-idae
Subfamilia	-oideae	-oideae	-inae
Tribus	-eae	-eae	(-ini)
Subtribus	-inae	-inae	

Codes resemble in all respects those of genera. They are uninomial and are singular nouns written with a capital initial letter.

Under the Botanical Code, several categories of such taxa are recognized (see Table 1). The name of such a taxon is not a uninomial but is a *combination* of the name of the genus[10] in which the taxon is classified with another term peculiar to the taxon and preceded by a word indicating its rank, e.g. *Costus* subg. *Metacostus*, *Primula* subg. *Primula* sect. *Primula* series *Acaules*. This term may be either a singular noun or a plural adjective and is written with a capital initial letter.

3.5 Names of species

The names of species consist of two terms and are therefore called *binomial*, *binominal* or *binary*. The name of a species consists of the name of the genus in which the species is classified followed by a second term which is peculiar to the species, e.g. *Equus caballus*, *Rosa acicularis*, *Corynebacterium fascians*.[11] The second term may be adjectival (in which case it must agree in gender with the generic name), a noun in apposition, or a noun (or rarely an adjective used as a noun) in the genitive case.[12] It is written with a small initial letter.[13]

Sometimes, especially in zoological literature, the name of the subgenus to which the species belongs may be written in parentheses between the generic name and the second term, e.g. *Anopheles* (*Myzomyia*) *gambiae*. However, this is not a part of the name of the species, which is always strictly binomial, and which in this instance is simply *Anopheles gambiae*.

The second term of the binary name of a species by itself has no standing and cannot be used alone to refer to any organism. For example, the scientific name of the oyster is *Ostrea edulis*. The oyster cannot be referred to simply as '*edulis*', as there are other species with this term as part of their names. Likewise, '*japonica*' by itself simply means Japanese and can refer to no particular kind of plant. On the other hand, in combination with various generic names it forms the names of various plant species, e.g. *Anemone japonica*, *Primula japonica*, *Chaenomeles japonica*.

Once the full name of a species has been cited in a text, its first term, i.e. the generic name, is often abbreviated to its initial letter in subsequent citations, if this can be done without causing ambiguity or doubt. Here, for example, we might now write *A. japonica*, as the full name of this *Anemone* species has already been mentioned and no ambiguity would be caused by so doing.

3.6 Names of taxa below the rank of species

The Zoological Code regulates the names of taxa of only one category below the rank of species—the subspecies. Names of taxa of infrasubspecific rank do not fall within its jurisdiction.[14] Names of subspecies consist of three terms and are therefore called *trinomial, trinominal* or *ternary*. The name of a subspecies consists of the name of the species in which it is classified followed by a third term which is peculiar to the subspecies, e.g. *Mus musculus domesticus*. Since in Zoology there is only one category of taxa to which such trinomials may be applied, they are written without a word indicative of rank preceding the third term.

The Bacteriological Code also excludes the names of taxa of infrasubspecific rank from its formal nomenclature, although it does provide certain recommendations which apply to them (see page 47). Names of subspecies under the Bacteriological Code have exactly the same form as under the Zoological Code, though a word indicating the rank is preferably inserted, e.g. *Bacillus subtilis* subsp. *niger* in preference to *Bacillus subtilis niger*. In contrast to the Botanical Code, the category names *subspecies* and *varietas* are alternative names for the same category (see Table 1); the latter is *not* subordinate to the former.

Under the Botanical Code, several categories of taxa below the rank of species are recognized (see Table 1). The name of such a taxon consists of the name of the species in which it is classified, followed by a term peculiar to the taxon preceded by a word indicative of rank, e.g. *Silene dioica* subsp. *zetlandica, Salix repens* var. *fusca*. Names of subspecies are therefore trinomial and have exactly the same form as under the Zoological and Bacteriological Codes.[15] Names of infrasubspecific taxa may consist of more than three terms, e.g. *Salix repens* subsp. *repens* var. *fusca*, but such names are long and clumsy and are reduced to ternary form unless this would cause ambiguity. For this reason, the insertion of words indicative of rank is obligatory under the Botanical Code.

The use of binary names for taxa below the rank of species is inadmissible under all three Codes.

3.7 Name groups

Under the Zoological Code, there are *name groups*. The names of a group are all co-ordinate nomenclaturally, that is, they are subject to the same Rules and Recommendations, irrespective of the rank of the taxa to which they apply. Names of taxa of the ranks of *Subtribus* to *Superfamilia* inclusive are *family-group names*. Names of genera and subgenera are *genus-group names*. Names of species and subspecies are *species-group names*.

Genus-group and species-group names are also co-ordinate under the Bacteriological Code. The Botanical Code recognizes no name groups but the names of taxa below the rank of species are in fact co-ordinate in some respects. The significance of name groups will become apparent when priority is discussed in Chapter 5.

3.8 Names, rank and position

The provisions of the Codes governing the forms of names have two important consequences. The name of a taxon indicates its rank, and, in the case of a species or taxon of lower rank, also its taxonomic position (the genus in which it is classified). In the next chapter the implications of these consequences are discussed.

4
Stability and Change

4.1 Instability of nomenclature

Since the name of a taxon indicates its rank and sometimes position, a change in rank and/or position may require a change of name. As a result, like common names, scientific names are by no means either unambiguous or universal. Many taxa have been known, simultaneously or successively, by two or more different names. This instability of biological nomenclature is a real disadvantage, for it means that, as a reference system, it cannot be said to be very efficient. Nevertheless, it is the only generally accepted system we have, and there are certain features of the classification of living organisms which inevitably cause difficulties in their nomenclature.

4.2 The sources of difficulty

The major source of difficulty is that our systems of classification are subject to continuous change. The better a system of classification reflects the various degrees of overall similarity between living organisms, the greater will be its predictive value, the more reliable a summary it will give of the properties of the organisms classified, the more convenient and comprehensive it will be in the storage of data and in the ease of recovery of information from it, and the more reliably it will lead to the correct placing of a new organism in the system. In other words, the better it will be. How well a classification reflects such overall similarity is tested every time our knowledge of the living world increases. As more organisms are discovered and as our knowledge of biological structure and function deepens, deficiencies of classification become more apparent and changes have to be made. The criteria by which taxa are defined are in consequence a *result* of the process of classification and cannot be stipulated *a priori*. In this respect taxa differ from the majority of scientific concepts, such as energy, work, molecule, operon, which are accorded an agreed, precise and invariable meaning. Moreover, not only do taxonomic criteria change with time (as knowledge increases) but also, at any one time, there may be differences of opinion between systematists as to how they should be employed. As a result, the circumscriptions of taxa are liable to change. Since this is as true for higher taxa as for lower taxa, the systematic positions of taxa within the system are likewise liable to change.

The *rank* assigned to a taxon at any one time is also liable to be changed. This is because there are no definitive criteria which can be used in the assignment of rank to a taxon. There is no way of defining a Class, for example, that is not equally applicable to a Subclass, an Order, a Family, or even to any taxon above the rank of species. Such taxa should be so circumscribed that all members of any one taxon resemble one another overall more closely than they resemble any member of any other taxon of the same rank. There should also be a comparatively greater discontinuity in variation between different taxa of the same rank than between the members of any one such taxon. But the only criteria for the assignment of rank are the opinions of the systematists who have studied the organisms concerned. Such opinions are affected by tradition, by the size of the taxa involved and by the degree of discontinuity between them. Nevertheless, the assignment of rank remains arbitrary and to a large extent subjective.

Most of the above applies equally to the delimitation of species and the assignment of specific rank. In theory at least, it is true, there are objective criteria by which specific rank can be assigned and specific limits set, at least in outbreeding sexually reproducing organisms.[16] But in practice these criteria are impossible to apply consistently. Many organisms reproduce wholly or almost wholly by asexual means. Amongst sexual organisms there are difficulties caused by inbreeding, hybridization and polyploidy. What is more, our knowledge of living organisms is still insufficient for adequate characterization of most species even in purely morphological terms. As a result, taxa at any one time considered as species are as likely to be changed in circumscription, position and rank as any other.

4.3 Consequences of systematic changes

A change in circumscription may involve either the union of two or more taxa previously considered distinct or the splitting of what was previously considered to be one taxon into two or more distinct taxa. In either case, at least one of the taxa involved will undergo a change of name. A change in the rank of a taxon will in many cases also cause its name (or at least the suffix indicative of rank) to be changed. A change in the systematic position of a taxon will also cause a change in its name if it is a taxon below the rank of genus (in Botany) or subgenus (in Bacteriology and Zoology). The most frequent, and the most upsetting, are changes in the names of species. Since the name of a species is a binary combination, the first part of which is the name of the genus in which the species is classified, the transfer of a species from one genus to another inevitably involves a change of name.

4.4 Advantages of the system

A system of nomenclature such as this, in which names are liable to change when changes in classification take place, has obvious defects as a reference system. They are offset, however, by certain advantages. Since the name indicates the rank and (sometimes) position of a taxon, it is a shorthand summary of the place of the taxon in the system of classification. This is not only an aid to memory but it also enables us to make inferences about organisms known to us only by name. Thus if we have a plant of *Rhododendron ponticum* in our garden, then merely by knowing the name *Rhododendron sikkimense* we may infer, without ever having seen them, that plants of the taxon to which this name applies will have a general overall resemblance to the plant in our garden. If names at all levels were independent of classification, these advantages would be lost. The mnemonic value of names would also be lost if they were replaced, as is sometimes advocated, by a system of reference numbers.

4.5 The conflict and its resolution

We are thus faced by an inherent conflict between the need for stability in nomenclature and the inevitable changeability of the classification it serves. The Codes of Nomenclature represent the cumulative results of years of effort to resolve the conflict, or rather, since it is unresolvable, to achieve a workable, regulated compromise. Their complexity is a measure of the difficulty of their task, but an understanding of the difficulty at least enables us to appreciate the provisions of the Codes which otherwise might seem needlessly complicated and cumbersome.

5

Operative Principles of Nomenclature

5.1 The aims of the codes

The Codes of Nomenclature try to ensure, that *with any given circumscription, position and rank*, a taxon can have one, and only one, name by which it may be properly known. This is the nearest approach to nomenclatural stability that can be achieved in systematics. They also try to avoid, or reject, the use of names likely to cause ambiguity or confusion. To achieve these aims, the Codes lay down certain provisions which must be followed in the giving of names to taxa and in the use of names. These provisions are based on a number of what may be called *operative principles*, of which the chief are *publication, typification and priority*.

The way in which these principles are employed to determine the name by which a taxon should properly be known is indicated in Table 3. Names must also be in accordance with the conventions of *formulation* already dealt with in Chapter 3. Those which are not in Latin form, or are otherwise inappropriate to the rank of the taxon concerned, are excluded from use. The requirement of *legitimacy* (§5.11, p. 23) is also operative, except under the Zoological Code.

5.2 Publication

Since the circumscriptions and definitions of taxa are liable to change, it is essential to be able to check back, when necessary, and find out what kind of organism the author of a name had in mind when he or she first used the name. For this reason the Codes require that some descriptive matter, available for consultation by others, should be associated with a name when it is first given to a taxon. Two basic conditions must be fulfilled before a properly formulated name can have any status in biological nomenclature; it must be published in a *medium* that conforms to the requirements of the appropriate Code, and must be accompanied by *written matter* conforming to the requirements of the Code.

If the first condition is satisfied, the name is regarded as *published* (by the Zoological Code) or *effectively published* (by the Botanical and Bacteriological Codes). Essentially the Codes require publication in works that are printed, reasonably permanent, and made generally available to the interested public. For example, the spoken word, microfilm made from

Table 3 The Nomenclatural Filter (For explanation see text)

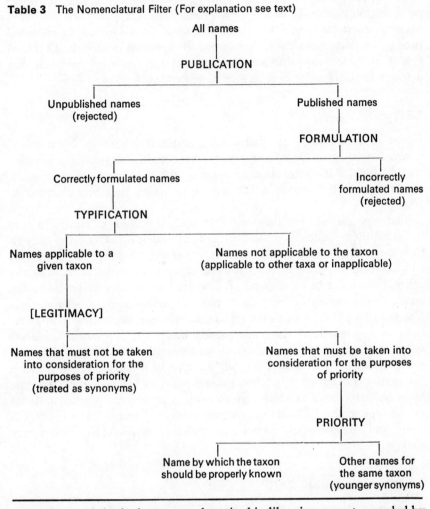

typescripts and single documents deposited in libraries are not regarded by the Codes as media of publication.

Even if these requirements are met, unless the name is accompanied by some information it will still be impossible to ascertain what kind of organism it was applied to. The Codes therefore require that certain information must accompany the original publication of a name. The requirements vary in detail from Code to Code and with the rank of the taxon concerned, but a general requirement is the provision of a *description* (or recognized equivalent) of the taxon to which the name is being given, or at least of some kind of reference to such a description. If this and the other pertinent requirements of the appropriate Code are met, then the name becomes *available* (under the Zoological Code), *valid* or *validly published* (under the Bacteriological Code) or *validly published* (under the Botanical Code). It thereby acquires nomenclatural status and must be taken into account for the purposes of biological nomenclature.

5.3 Typification

Under all three Codes, the application of names is determined by means of *nomenclatural types*. *Typification* is the process of designating a nomenclatural type. Publication is the means by which names enter biological nomenclature; typification is the means by which they are allocated to taxa.

The type method is fundamental to the application of names to taxa under all three Codes. The Zoological and Botanical Codes differ only in their conception of the way in which it acts as a link between nomenclature and classification. The Zoological Code regards the nominal taxon as the concept common to all notions of a taxon, the nominal taxon being the taxon, as objectively defined by its type, to which a given name applies. The Botanical Code regards the type as the concept common to all applications of a given name. In consequence, under the Zoological Code we speak of 'the type of a taxon' (as we do under the Bacteriological Code), while under the Botanical Code, of 'the type of a name'. The difference however is conceptual only; throughout biological nomenclature, the type is the objective basis to which a given name is permanently linked and it is by the type method that the correct application of names to taxa is objectively and unequivocally determined however much classification may change.

5.4 What is a type?

A type is an element on which the description associated with the original publication of a name was based, or is considered to have been based. The term 'element' here means different things according to the rank of the taxon concerned. Under the Zoological Code, the type of a family-group taxon is a genus, the type of a genus-group taxon is a species, and the type of a species-group taxon is a specimen. Under the Botanical Code, the type of the name of a family or lower taxon above the rank of genus is a genus, the type of the name of a genus or lower taxon above the rank of species is a species and the type of a name of a species or infraspecific taxon is a specimen (or sometimes a description or an illustration of a specimen). Under the Bacteriological Code, the type of a class or subclass is an order, the type of an order, suborder, family, subfamily, tribe or subtribe is the genus on the name of which the name of the higher taxon is based, the type of the name of a genus or subgenus is a species, and the type of a species or subspecies is preferably a living strain but may be a preserved specimen or preparation, an illustration or a description.

A type is purely a nomenclatural concept, and has no significance for classification. For example, specimens that are types are merely those which happen to have had names associated with them, and for the purposes of classification are treated like any others. As a result, a type falling within the range of variation of a taxon may stand at one extreme of that range. Nevertheless, the name to which that type is linked will apply to the taxon and may well be the name by which it should be properly known. In other words, the nomenclatural type associated with the name by which a taxon is properly known is not necessarily typical of the taxon in terms of range of variation. It is not the purpose of a type to be typical in the variational sense; the purpose of a type is to provide a fixed point associated with a name in the range of variation of organisms so that no matter where discontinuities are found to occur and boundaries between taxa drawn, the application of the name can be objectively and unequivocally decided.

5.5 How the type method works

The operation of the type method in deciding the application of names can be illustrated by considering the pine genus *Pinus*. When Linnaeus first published the name in 1753, he included within the genus the following species: *P. sylvestris, P. pinea, P. taeda, P. cembra, P. strobus, P. cedrus, P. larix, P. picea, P. balsamea* and *P. abies*. Later, with increased knowledge, it became apparent that Linnaeus's concept of the genus was far too wide, and that his species were better classified into five distinct genera, as follows:

Genus 1: *P. cedrus*
Genus 2: *P. larix*
Genus 3: *P. picea, P. balsamea*
Genus 4: *P. abies*
Genus 5: *P. sylvestris, P. pinea, P. cembra, P. strobus, P. taeda*

Classification being completed, nomenclature can now be considered. To which of these five genera is the name *Pinus* to be applied? The type method requires that it be the one in which the type of the name *Pinus* falls. The type of a generic name is a species, and in the case of the name *Pinus* it is the species *P. sylvestris*. This species falls into genus 5, and it is to this genus that the name *Pinus* must be applied. The other four genera must therefore be known by other names (which are, respectively, *Cedrus, Larix, Abies* and *Picea*).

5.6 Kinds of types

The Codes recognize several kinds of types, of which the following are the more important. A *holotype* is either the sole element used by the author of a name or the one element designated by him as the type. A *syntype* is either any one of two or more elements used by the author of a name who did not designate a holotype or any one of two or more elements designated by him simultaneously as types. A *lectotype* is an element selected subsequently from amongst syntypes to serve as the nomenclatural type. The designation of a lectotype must be based on careful consideration of all the evidence provided by the author of a name in the place of original publication. Each Code provides guidance as to the procedure that must be followed. A *neotype* is an element selected to serve as the nomenclatural type when through loss or destruction no holotype, lectotype or syntype exists. In the selection of neotypes similar care is needed and guidance is again given by the Codes. In bacteriology, where the type of a species or subspecies is preferably a living culture, many types are of necessity neotypes, although many so-called 'type cultures' are not in fact types in the nomenclatural sense at all.

If it proves impossible to typify a name satisfactorily, either through lack of information or because the type has been lost or is a mixture of discordant elements, then obviously the name cannot be applied to any taxon.

5.7 Priority

If two or more types fall within the range of variation of a taxon then there will be as many names that apply to the taxon, and some way of deciding by which it should be known will be necessary. The decision is made according to *priority*. The principle of priority requires that *when two or more names apply to the same taxon*, in general *it is by the oldest one that it should properly be known*. By oldest is meant the oldest available (Zoological Code) or the first validly published (Botanical and Bacteriological Codes). The name by which a taxon is properly known is called its *correct name* by the

Botanical and Bacteriological Codes and its *valid name* by the Zoological Code.[17] In the case of a taxon of the rank of species and below,[18] the term of the name peculiar to the taxon dates from the place of its original publication, irrespective of the combination in which it was originally published. Thus the name *Raphidiocystis chrysocoma* although first published in 1962 as such, is the name by which the taxon to which it refers is properly known, in spite of the existence of the name *R. welwitschii*, applicable to the same taxon and published in 1871, because the term *chrysocoma* was originally published in the combination *Cucumis chrysocomus* in 1827.

5.8 Limitations of priority

Certain limitations are set to the operation of the principle of priority. They include starting-point dates, limitations associated with rank, the exclusion of certain classes of names from consideration for the purposes of priority, and procedures for the conservation and rejection of names.

5.9 Starting-point dates

A starting-point date is the date of publication of a work previous to which no name is considered to have been made available (Zoological Code) or validly published (Botanical and Bacteriological Codes). Different groups of organisms have different starting-points, depending upon which systematic work is considered to have laid the foundation of the modern nomenclature of the group concerned. The following are the main starting-point works and the dates on which they are treated as having been published. Linnaeus, *Systema Naturae*, edition 10, 1 January 1758: *Animalia*. Linnaeus, *Species Plantarum*, edition 1, 1 May 1753: recent *Spermatophyta*, *Pteridophyta*, *Hepaticae*, *Sphagnaceae*, *Lichenes*, *Myxomycetes*, and most *Algae*. Sternberg, *Flora der Vorwelt*, *Versuch* 1:1–24, t. 1–13, 31 December 1820: fossil plants, all groups. Recent *Fungi*, other *Musci*, *Nostocaceae*, *Desmidiaceae* and *Oedogoniaceae* have their own starting-points, for details of which the Botanical Code should be consulted.

For *Bacteria*, the starting-point date as from 1 January 1980 will be 1 January 1980. Up to then, the starting-point date is 1 May 1753 (Linnaeus, *Species Plantarum*, edition 1). Names of *Bacteria* validly published up to 31 December 1977 will be considered by the Judicial Committee and lists of names will be prepared for approval by the International Committee on Systematic Bacteriology. When approved by the International Committee on Systematic Bacteriology, these Approved Lists of Bacterial Names will be published in the International Journal of

Systematic Bacteriology. Names validly published under the Bacteriological Code between 1 January 1978 and 1 January 1980 will be added to these Approved Lists of Bacterial Names. After 1 January 1980, no further names will be added to the Lists. Those names validly published prior to 1 January 1980 but not included in the Approved Lists will have no further standing in bacterial nomenclature, and will be treated as if they did not exist. If two names compete for priority and if both names date from 1 January 1980 on an Approved List, the priority will be determined by the date of original publication of the name before 1 January 1980.

5.10 Limitations of priority with respect to rank

Under the Botanical Code, priority does not apply to names of taxa above the rank of family. For taxa of family rank or below, priority is restricted to within each rank and in no case does a name have priority outside the rank of the taxon to which it applies. Thus the taxon *Campanula* sect. *Campanopsis* (1810) when raised to generic rank must be called *Wahlenbergia* (1821) which is the earliest name for the taxon at generic rank.

Under the Zoological Code, priority is not so severely restricted with respect to rank. Priority is operative within each of the three name-groups (see p. 11) irrespective of difference in rank within the group. Thus the name given to a taxon within, say, the family-group is available from its original date of publication at any rank within the Family-group irrespective of the rank of the taxon to which it was applied when it was first published. For species-group names and genus-group names the same applies.

The Bacteriological Code is intermediate in this respect between the Botanical and Zoological Codes. Specific and subspecific names, and generic and subgeneric names, form two name-groups corresponding respectively to the species-group and genus-group of the Zoological Code, and within each group priority is likewise operative irrespective of differences in rank. On the other hand, there is nothing corresponding to the family-group and for taxa above the rank of genus priority is restricted, as under the Botanical Code, to within each rank. Under the Bacteriological Code, priority does not apply to names of taxa above the rank of order.

5.11 Names excluded from consideration for the purposes of priority when the name by which a taxon should properly be known is being decided

There are several kinds of names which although available (Zoological Code) or validly published (Bacteriological and Botanical Codes) are excluded from consideration for the purposes of priority when the name by which a taxon should properly be known is being decided. The exclusion may be absolute or may operate only under certain circumstances as prescribed by the appropriate Code. Names that are not in accordance with the provisions of the Code such that they must not be taken into consideration for the purposes of priority when the correct name of a taxon is being decided are termed *illegitimate* by the Botanical and Bacteriological Codes,[19] but this term is not employed by the Zoological Code, which, unlike the Bacteriological and Botanical Codes, does not make use of the concepts of legitimacy and illegitimacy.

The most important of such names are names which are later (junior) homonyns, which must be rejected under all three Codes. Others include certain genus-group names ending in *-ites*, *-ytes* or *-ithes* and given only to fossils (Zoological Code), nomenclaturally superfluous names (Botanical and Bacteriological Codes), tautonyms (Botanical Code), names of fossil taxa (except algae) when in competition with names of recent taxa (Botanical Code), names the types of which are imperfect states of fungi (see §8.3, p. 40) when in competition with names the types of which are perfect states (Botanical Code), and widely and persistently misapplied names (Botanical Code). Also not to be used are names rejected under the procedure for the conservation and/or rejection of names provided for by the appropriate Code (see §5.16 p. 24).

5.12 Homonyms

Homonyms are names spelt in an identical manner[20] but based on different types. Clearly, confusion would result if such names came into widespread use; the need for unambiguity in scientific names would not be met. The Codes therefore rule that of two or more homonyms, all except the oldest are excluded from use.[21] Later (or junior) homonyms can therefore never be names by which taxa can properly be known and the possibility of confusion resulting from the same name meaning different taxa to different people is thus minimized.

Under the Bacteriological Code, a name must be rejected if it is a later homonym of the name of a taxon of bacteria, algae, fungi, viruses or protozoa, i.e. if it duplicates a name previously validly published for a taxon of the same rank and based on a different type.[22] The Botanical Code rules similarly, except that names of animal taxa need be considered only if the taxa were once included in the plant kingdom; otherwise the names of plants and animals are independent. The Zoological Code likewise excludes from homonymy names that have never been used for taxa in the animal kingdom. It defines homonymy as the identity in spelling of names based on different types within a genus, the genus-group or the family-group. Family-group names differing only in suffix are also considered to be homonyms. Unlike the Botanical and Bacteriological Codes, the Zoological Code explicitly states that two identical species-group names placed in different genera that have homonymous names are not to be considered as homonyms. Thus *Noctua variegata* (*Insecta*) and *Noctua variegata* (*Aves*) are not to be considered as homonyms.

5.13 Superfluous names

A name is nomenclaturally superfluous when published (*nomen superfluum*) if the taxon to which it was applied, as circumscribed by its author, included the type of another name which ought to have been adopted under the rules. The details of this Rule and its interpretation are matters of some complexity and are beyond the scope of this Handbook. The appropriate Code (Bacteriological or Botanical) should be consulted. The purpose of this rule is to prevent the needless multiplication of names.

5.14 Tautonyms

A *tautonym* is a name of a species in which the second term exactly repeats the generic name, e.g. *Bison bison*. Tautonyms are illegitimate under the Botanical Code.[23] In contrast, the Zoological and Bacteriological Codes permit the use of tautonyms.

5.15 Widely and persistently misapplied names

Under the Botanical Code, a name must be rejected if it has been widely and persistently misapplied, i.e. used for a taxon not including its type. Names thus rejected are to be placed on a list of rejected names.

5.16 Conservation and rejection of names

In order to promote stability and continuity in nomenclature, all three

Codes provide for the making of exceptions to the Rules, so that disadvantageous changes that would be caused by their strict application can be avoided.

Under the Zoological Code, the International Commission on Zoological Nomenclature has power to suspend the application of any provision of the Code, to suppress or validate any name, and to annul or validate any typification, publication or any published nomenclatural act. The decision of the Commission on any particular case referred to it is termed an *opinion*. Opinions are published by the International Trust for Zoological Nomenclature and become operative on publication. The Trust also publishes *declarations*, i.e. provisional modifications of the Code the Commission is empowered to make between Congresses, the official indexes of rejected and invalid names and works, and the official lists of validated names and approved works. Although published separately, each instalment of an official list or index is considered to be an integral part of the Zoological Code.

The Bacteriological Code also provides for the rejection and retention of names. Names to be retained are called *nomina conservanda*, those to be rejected, *nomina rejicienda*. Proposals for the conservation and rejection of names must be submitted to the Judicial Commission which gives an opinion upon each proposal. The Judicial Commission can also issue opinions relative to the interpretation of any of the provisions of the Code if so requested. The opinions of the Judicial Commission become operative unless rescinded by a majority vote of the International Committee. Only the Judicial Commission can place names on the lists of conserved and rejected names. Provision is made by the Code for the rejection, amongst others, of ambiguous names (*nomina ambigua*), doubtful names (*nomina dubia*), names causing confusion (*nomina confusa*) and perplexing names (*nomina perplexa*). Definitions of these will be found in the glossary/index. Currently conserved family and generic names, conserved specific epithets, rejected generic and subgeneric names and rejected specific epithets are listed in an appendix to the Code.

The circumstances under which exceptions may be made to the Botanical Code are much more restricted. It is possible only to conserve or reject names of taxa of the ranks of genus to family inclusive.[24] Conservation or rejection of the names of species is not permitted.[25] Proposals for the conservation or rejection of names must be submitted to the General Committee on nomenclature for study and approval by the appropriate Special Committee. If approved, they are submitted to an International Botanical Congress for adoption. Appendices to the Code list currently conserved and rejected names.

5.17 Orthographic variants

All three Codes give, in more or less detail, rules and recommendations according to which names must be spelt and transliterations made into biological Latin from other languages. They cannot be considered in detail here, and the Codes should be consulted by the interested reader. It is sufficient for the non-systematist to be aware that two or more orthographic variants—different spellings—of the same name may exist, by only one of which can the taxon be properly known. Such variants are considered to be forms of the same name[26] and if one such variant is a later homonym, none of the others may be used in its stead.

6
Name-changes and Synonymy

6.1 Name-changes

Name-changes not only tend to annoy those who are affected by them but also reduce the efficiency of biological nomenclature as a reference system. To reduce them to a minimum, the Codes of Nomenclature precisely specify the circumstances under which a name must be changed, and in what way. The alteration of names is otherwise not permitted. Under all three Codes, the name of a taxon may not be changed merely because someone happens to think it inappropriate or objectionable, or because another is considered better known, or because it has lost its original meaning. Thus the name *Scilla peruviana* may not be rejected merely because the species to which it refers does not occur in Peru. This is in accordance with the basic principle that a name is primarily an arbitrary symbol the purpose of which is to facilitate communication. A change in the name by which a taxon has become known is permitted by the Codes only if it is necessitated by a correction of nomenclatural error, by a change in classification or by a correction of a past misidentification.

6.2 Nomenclatural reasons

A name that is in common use may have to be changed for nomenclatural reasons, i.e. because it is not in accordance with the requirements of the appropriate Code. Thus the name *Viburnum fragrans* (published in 1831) by which a commonly cultivated shrub became widely known had to be replaced by *V. farreri* (1966), in consequence of its being a later homonym of a *V. fragrans* published in 1824 by another botanist for a different species. There are three main causes of such purely nomenclatural changes. First, names have often in the past come into use instead of those which should have been adopted under the present Codes. This is mainly because although the Codes are modern most of their provisions are retroactive; unfortunately, there are also some contemporary workers who deliberately or through ignorance do not observe the requirements of the Codes. Secondly, many names have in the past been misapplied, usually through lack of proper typification; only in the 20th century has the type concept been fully developed. Thirdly, many names have come into use in violation of the principle of priority because earlier names were published in more or less obscure works and overlooked by subsequent authors.

Name-changes for nomenclatural reasons have been particularly trouble-some in recent years as systematists have endeavoured to bring nomen-clature into line with the requirements of the International Codes. How-ever, the bringing to light of overlooked names in the old literature is perhaps nearing completion. Together with a sustained effort by system-atists to achieve general agreement in the typification and application of all names and a strict adherence by all workers to the provisions of the Codes, it is hoped this will lead to name-changes for nomenclatural reasons becoming ever fewer and fewer until eventually they cease to trouble us.

6.3 Taxonomic reasons

Unfortunately, the same cannot be said of name-changes which become necessary for taxonomic reasons. These arise from taxonomic research itself and are inevitable accompaniments of our systems of classification which, as was explained in Chapter 4, are constantly being modified as our knowledge of living organisms increases. The Codes do not, of course, permit the name of a taxon to be changed merely because its diagnostic characters are altered or its circumscription changed. Only if such modi-fications involve a change in taxonomic position and/or rank, or union with another taxon, may a name-change become necessary under the provisions of the appropriate Code (see p. 14).

6.4 Synonyms and synonymy

Two or more names that are considered to apply to the same taxon are known as *synonyms*. Of a number of synonyms, therefore, according to the principle of priority, only one can be the name by which the taxon may be properly known—in general, the oldest (senior) one. The later (or junior) synonyms then form what is called the *synonymy* of the accepted name of the taxon. It is important in the consultation of taxonomic works clearly to distinguish the names accepted as correct (or valid) from those cited in synonymy. They are usually distinguished typographically; synonyms may also be indicated by being preceded by the abbreviation 'syn'. It is, unfortunately, not always as clearly indicated as it might be which name is the one to be used and which form the synonymy; care on the part of the user is needed.

Modern taxonomic research reduces many names that have previously been held to apply to different species to synonymy. This frequent excess of names over taxa has come about in two main ways—through lack of

awareness of previously published names, or through insufficient appreciation of the amount of variation that can exist within a species. Mere variants
or races of one species have been given different names at specific rank.
This was often a result of lack of sufficient specimens, especially of tropical
organisms. Nowadays, with more material available and greater oppor-
tunities for field and experimental studies, there is better appreciation of
the limits of species. Modern communications and international taxonomic
associations also reduce the likelihood of the same taxon being described
more than once under different names, although keeping abreast of the
current literature is still a problem in spite of the advent of computerized
abstracting and data-handling services.

6.5 Taxonomic and nomenclatural synonyms

There are two kinds of synonyms, taxonomic and nomenclatural. *Nomenclatural synonyms* are synonyms based upon the same type. Their synonymy
is therefore absolute, not a matter of taxonomic opinion. Hence they are
also known as *obligate, objective* or *homotypic* synonyms. *Taxonomic
synonyms*, on the other hand, are synonyms based upon different types, and
remain synonyms only as long as their respective types are considered to
belong to the same taxon. They are therefore also known as *subjective* or
heterotypic synonyms. Nomenclatural synonymy may be indicated by use
of the mathematical sign of congruence, $\equiv$; taxonomic synonymy, by the
sign of equality, $=$.

6.6 The significance of synonymy

Although the names that an author places in synonymy are not correct
(Botanical and Bacteriological Codes) or valid (Zoological Code), this does
not imply they are of no significance. A considerable amount of information
may be recorded in the literature under one or more of these names. The
synonymy of a taxon, therefore, is a key to information about the taxon. It
is for this reason that taxonomic research is concerned, among other things,
with the correct establishment of synonymies. The establishment of a
synonymy represents a synthesis of our knowledge of the organisms
concerned.

6.7 Nomenclature and classification

The International Codes are so formulated that in any given classification,
a taxon can have only one name by which it may properly be known. The

qualification 'in any given classification' is important, for it allows for flexibility in nomenclature when changes are made in classification. The populations of the genus *Raphidiocystis* occurring in Madagascar were considered by Baker in 1890 to belong to two distinct species, which he called *R. brachypoda* (1882) and *R. sakalavensis* (1890). Under Baker's classification, therefore, both these names are correct, *R. brachypoda* for one species, *R. sakalavensis* for the other. On the other hand, Jeffrey and Keraudren in 1967 considered the Madagascan populations all to fall within the limits of a single species. Under their classification, which treats *R. brachypoda* and *R. sakalavensis* as synonyms, only the former name is correct. The latter, if used for the species as circumscribed by these authors, would be incorrect under the Botanical Code. Another example of the dependence of names on classification is afforded by the case of the zonal pelargoniums—the 'geraniums' of the gardener. If like most contemporary botanists, we recognize the genus *Pelargonium* as taxonomically distinct from *Geranium*, then *Pelargonium zonale* is the correct name for the species to which these plants belong and the name *Geranium zonale*, given to it by Linnaeus in 1753, is incorrect and a synonym. But if we so wished we could follow the older classification of Linnaeus under which the two genera were united. If we did this, then *Geranium zonale* would be the correct name for the species and *Pelargonium zonale* incorrect and a synonym. These examples serve to emphasize that the name by which a taxon should properly be known is determined by the classification adopted as well as by the requirements of the appropriate Code of Nomenclature. Unless a given classification is specified, it is meaningless to ask what is the correct (under the Botanical and Bacteriological Codes) or valid (under the Zoological Code) name of a taxon. The names usually given by systematists in reply to enquiries by others are those correct (or valid) under the currently generally accepted classification of the group of organisms concerned.

6.8 Misidentifications

It sometimes happens that an organism which has become well-known under a certain name is later found to have been misidentified. The name by which it has become known really applies to a different organism. For example, an African species of the orchid *Polystachya* was in 1929 identified as *P. obanensis* and was referred to in the published literature under this name until 1960 when it was shown not to be identical with the true *P. obanensis*. It was in fact a new, undescribed and un-named species. The name *P. bella* was then published for it. As a result, those who had known it as *P. obanensis* had to get used to the fact that it had merely been misidentified as such and was really *P. bella*. Since errors of identification are always possible, such name-changes due to mis-identification will occur from time to time, but they are unlikely to be numerous.

7
Authorities and their Citation

7.1 Authorities

The scientific names of organisms are often written followed by one or more personal names, sometimes abbreviated, e.g. *Primula* L., *Primula vulgaris* Huds., *Canis lestes* Merriam, *Serratia marcescens* Bizio. These personal names are the *authority citations* for the names they follow. The *authority* of a scientific name is the name of the author of the name. The author of a name is the person who first published the name in a way that satisfies the criteria of valid publication (under the Botanical and Bacteriological Codes) or of availability (under the Zoological Code). Thus the citation

Primula vulgaris Huds.

tells us that Hudson was the first person validly to publish the name *Primula vulgaris*.

The authority citation is not part of a scientific name but is essentially an abbreviated bibliographic reference the use of which adds to nomenclatural precision. Authority citations are especially helpful in the distinguishing of homonyms, e.g. *Viburnum fragrans* Bunge (1831), a later homonym of *V. fragrans* Loisel (1824); *Disticta* Wasmann, 1916 (*Coleoptera*), a junior generic homonym of *Disticta* Hampson, 1902 (*Lepidoptera*).

7.2 The use of *in* and *ex*

Sometimes authority citations consist of two names connected by the Latin preposition *in* or *ex*. A citation such as

Viburnum ternatum Rehder in Sargent

means that Rehder was responsible for the valid publication of the name *V. ternatum* in a work edited (or otherwise written) by Sargent. In such cases it is the name *before* the *in* which is the name of the publishing author, and if it is necessary to use a shortened form of the citation, the 'in Sargent' part may be omitted. It is added solely to help the location of the publication concerned in bibliographies and library catalogues.

Under the Botanical Code, a citation such as

Gossypium tomentosun Nutt. ex Seem.

means that Seeman was responsible for the valid publication of the name *Gossypium tomentosun*, which had originally been coined by Nuttall, but which he himself had never validly published. He may merely have written the name on the label of a specimen, or in a manuscript, or he may have effectively published it without satisfying the criteria of valid publication, e.g. by failing to provide a description of the plant to which he applied it. In such cases it is the name *after* the *ex* which is the name of the publishing author, and it is the name before the *ex* which may be omitted if it is necessary to shorten the citation. Thus for the example given, it is perfectly correct to cite just *Gossypium tomentosum* Seem.

Under the Bacteriological and Zoological Codes, however, the name of the publishing author is cited *before* the *ex*; e.g. under the Bacteriological Code, the citation

Bacillus caryocyaneus Dupaix ex Beijerinck

means that Dupaix was responsible for the valid publication of a name originally coined by Beijerinck, but which he himself never validly published.[27]

7.3 The use of parentheses and the double citation

Under the Zoological Code, if a species-group taxon first described in one genus is later transferred to another, the name of the original author is cited enclosed in parentheses when the new combination is cited. Thus, the species first called *Staphylinum discoideus* Gravenhorst was later transferred to the genus *Philonthus*, and the authority when it is referred to this genus is cited as follows: *Philonthus discoideus* (Gravenhorst). If it is desired to cite the name of both the original author of a species-group name and the person who transferred the taxon to which it applies to another genus, the name of the latter may be placed after the parentheses that enclose the name of the original author, e.g. *Philonthus discoideus* (Gravenhorst) Nordmann.

The use of this so-called *double citation* is purely optional under the Zoological Code. Moreover, mere changes of rank within the genus do not require a change in the authority citation. On the other hand, the Botanical and Bacteriological Codes require the use of the double citation, not only when a taxon below the rank of genus is transferred to another taxon, but also when a genus or taxon of lower rank is altered in rank but retains its original epithet. Under the Botanical and Bacteriological Codes, therefore, a double citation indicates there has been a change in taxonomic position and/or rank.

An example of a change in taxonomic position is afforded by the names of the species known in English as the hollyhock. This was originally placed by Linnaeus in the mallow genus as *Malva rosea* L. Later, Cavanilles decided it was better placed in a distinct genus *Althea*, and validly published the new combination required. This taxonomic position is accepted today, so we now know the hollyhock as *Althea rosea* (L.) Cav. Likewise, *Treponema pallidum* (Schaudinn et Hoffmann) Schaudinn was originally described as *Spirochaete pallida* Schaudinn and Hoffmann and later transferred to the genus *Treponema* by Schaudinn, who validly published the required new combination *T. pallidum* (Schaudinn et Hoffmann) Schaudinn.

A botanical example of a change in taxonomic rank is given by *Populus canescens* (Aiton) Smith. In 1789 Aiton described a new variety of the species *Populus alba*, for which he validly published the name *P. alba* var. *canescens* Aiton. In 1804 Smith showed this in fact to be a distinct species and raised Aiton's variety to specific rank, validly publishing for it the name *Populus canescens* (Aiton) Smith.

The significance of the double citation is that it shows that the type of the newer name (published by the author after the parentheses) is the same as the type of the older name (published by the author within the parentheses). In other words, it indicates that the newer name is based upon the older one. For this reason, the older name is often called the *basionym* of the newer one. Thus, *Malva rosea* L. is the basionym of *Althea rosea* (L.) Cav. In the case of a change in taxonomic position without a change in rank, the double citation also indicates that the epithet concerned dates for the purposes of priority from the date of valid publication of the basionym, not from the date of valid publication of the new combination. Thus *Raphidiocystis chryscoma* (Schumach.) C. Jeffrey (1962) has priority over its synonym *R. welwitschii* J. D. Hooker (1871) since its basionym *Cucumis chrysocomus* Schumach. (1827) is older; the specific epithet *chrysocomus* dates from 1827.

7.4 The use of square brackets

Under the Zoological Code, the enclosure of an authority citation in square brackets indicates that the name of the taxon was published anonymously. In such cases, the authority citation (within the square brackets) has obviously been ascertained indirectly; the use of square brackets denotes the original anonymity.

Under the Botanical Code, a citation in square brackets indicates a pre-starting-point author. For example, the generic name *Lupinus* L. (1753) had previously been effectively published in 1719 by the pre-starting-point author Tournefort and its authority may therefore be cited as *Lupinus* [Tourn.] L.

In citations of synonymy, square brackets are often used to enclose statements of misidentification, see §7.7, p. 36.

7.5 The citation of dates

The date of a name is the date of publication of the work in which the criteria of valid publication (Botanical and Bacteriological Codes) or availability (Zoological Code) were first satisfied. In the absence of evidence to the contrary, the date given in the work concerned must be accepted as correct. In many cases, however, the dates are not correct and for the proper establishment of priority in taxonomic research it is essential to check, if possible, the dates of publication from all available internal and external evidence. If this is not done, disadvantageous name-changes may result from further more thorough bibliographic researches.

Dates of publication are usually cited either with the authority citation or as part of the full bibliographic reference. Under the Zoological Code, the date of publication must follow the authority citation with a comma interposed, thus: *Staphylinus maxillosus* Linnaeus, 1758. The date should be enclosed in parentheses only if it is established from internal evidence other than a specific statement of the date of publication in the work concerned. If the date of publication is established solely from external evidence (e.g. notices in other works or publisher's correspondence), it should be enclosed in square brackets.

Different procedures are required by the Botanical and Bacteriological Codes. Under the latter, the date follows the authority citation directly without a mark of punctuation interposed, thus, *Serratia marcescens* Bizio 1823. The Botanical Code recommends that the authority name, if unabbreviated, be separated from what follows by a comma; abbreviated names need no punctuation other than the period (full stop) indicating abbreviation. Dates may be enclosed in parentheses, and this is often done in botanical literature, e.g. *Trochomeria* Hook. f. (1867).

7.6 The citation of homonyms

Homonyms may be distinguished by the citation of their authorities and dates, and, if necessary, by the use of the conjunctions *non* ('not of') and *nec* ('nor of'). The citation

Pfeifferella Labbe 1899 non Buchanan 1918

indicates it is the taxon described under the name *Pfeifferella* by Labbé in 1899 that is being referred to, and not the taxon to which Buchanan in 1918 gave the same name. The citation

Bartlingia Brogn. (1827) non Reichb. (1824) nec F. v. Muell. (1877)

makes it clear that it is the taxon described as *Bartlingia* by Brogniart in 1827 that is being referred to, not the earlier and later homonymous taxa described respectively by Reichenbach and F. von Müller.

7.7 The citation of subsequent usages, misidentifications and the use of *sensu*

The citation of an authority together with a bibliographic reference and date has a very precise meaning in taxonomic literature. It indicates that the person whose name is cited first published the name concerned, in the work cited and on the date cited, in a way that satisfies the criteria of valid publication (under the Botanical and Bacteriological Codes) or of availability (under the Zoological Code). Later, of course, other authors in other literature make use of the same name to refer to specimens or populations which they consider to belong to the same taxon as those studied by the original author. When it becomes necessary to refer to such subsequent users of the name, it is important that they be cited in a way that does not give the impression that they are authors of later (junior) homonyms.

The distinction is usually made typographically, by the use of a punctuation mark other than a comma or full stop, e.g. a colon or a semicolon, as in the citation:

Diplocyclos decipiens (Hook. f.) C. Jeffrey (1962); R. and A. Fernandes, Consp. Fl. Angol. 4:279 (1970)

This shows that R. and A. Fernandes, in their 1970 publication, made use of the name validly published for the first time by C. Jeffrey in 1962, to refer to what they considered were plants conspecific with those studied by Jeffrey.

Later usages are cited in this way if it is certain that the subsequent authors employed the name in the correct sense, i.e. applied it to organisms, the taxonomic identity of which with the type of the name concerned there is no reason to doubt. On the other hand, many names have been used in more than one sense, i.e. have been applied by different authors to taxonomically different, although superficially similar, organisms. This situation has come about mainly because the descriptions provided by the original authors of the names were too brief and vague to permit unequivocal typification. For example, the fungus depicted by Wakefield and Dennis in their 1950 book on British Fungi under the name *Mycena filopes* ([Bull. ex] Fr.) Quel. is certainly that species as understood by Jakob Lange in his Danish agaric flora (1936) but it is certainly not the species to which Kühner in his monograph of the genus *Mycena* (1938) applied this name. Nor is it certain to which taxon this name and its basionym, *Agaricus filopes* [Bull. ex] Fr., published by Fries in 1821, should be applied. In other words, it is a name of doubtful application which has also become an ambiguous name by being subsequently used by different authors in different senses. The use of a name in ways which may or may not be different from its original application can be indicated by means of the term *sensu* ('in the sense of') in citations. Thus we may cite:

Mycena filopes ([Bull. ex] Fr.) Quel. sensu Lange (1936); sensu Wakefield et Dennis (1950); non sensu Kuehner (1938)

This tells us that Wakefield and Dennis are using Quélet's name (itself based on an earlier name first effectively published by the pre-startingpoint author Bulliard and later first validly published by Fries) in the sense in which it was used by Lange, but not in the sense in which it was used by Kühner.

The construction *sensu . . . non* can also be used to cite a misidentification. An example is afforded by the case of the orchid species *Polystachya bella* already considered (p. 30). While this species was misidentified as *P. obanensis* it was several times referred to in the literature under this name. When publications in which this occurred are cited, it must be made clear that the name as used refers to *P. bella* and not to the true *P. obanensis* Rendle. This can be done as follows:

P. bella Summerh. in Kew Bull. 14:137 (1960). Syn.: *P. obanensis* sensu Moreau in Journ. E. Afr. Nat. Hist. Soc. 17:30 (1947) et sensu Hey in Orch. Rev. 69:278 (1961), non Rendle

In this way it is made clear that the true *P. obanensis* Rendle is not conspecific with *P. bella* Summerh. but is a different species.

When they are included in lists of synonymy, statements of misidentification are commonly enclosed in square brackets.

It is important to note that it would be incorrect to cite *P. obanensis* Moreau non Rendle instead of *P. obanensis* sensu Moreau non Rendle. The former kind of citation would imply (as we have seen) that Moreau himself had described a new species for which he validly published the name *P. obanensis* Moreau, which would then have been a later homonym of *P. obanensis* Rendle. This, of course, he did not do; he merely wrongly identified the plant he was referring to as *P. obanensis* Rendle and had no intention of publishing a new name. The use of *sensu . . . non* in citation makes this clear.

Misidentification can also be cited by the use of the phrase *auct.* (of authors) or *sensu auct.* (in the sense of authors) with *non* as in the following construction: *P. bella* Summerh.; *P. obanensis* auct. non Rendle: Moreau in Journ. E. Afr. Nat. Hist. Soc. 17:30 (1947); Hey in Orch. Rev. 69: 278 (1961).

This section must end with a word of warning. The above procedures of citation are those which should, ideally, be used. They are not, however, always those that have been used in the appropriate situation. In the literature one may find, therefore, misidentifications cited so that they appear to be later homonyms, or later applications so that they might be mistaken for authority citations. Even authorities themselves may be erroneously cited; names may be misascribed to later authors or even to individuals that discovered taxa, not to those who first published their names in accordance with the requirements of the appropriate Code. Users of taxonomic literature must be prepared to meet such pitfalls.

7.8 The use of qualifying phrases

An alteration in the diagnostic characters or circumscription of a taxon does not warrant a change in the authority citation for the name of the taxon. However, if necessary, a suitable qualifying phrase may be added, after the authority citation, and followed by the name of the author responsible for the alteration, e.g. *Taenia solium* Linnaeus, partim Goeze, *Globularia cordifolia* L. excl. var. (emend. Lam.), *Bacillus* Cohn emend. Migula. Other examples of such qualifying phrases will be found in the Glossary-Index (p. 55).

If an author emended a taxon so as to exclude from its circumscription the type of its name, which nevertheless he continued to apply to the emended taxon, such qualifying phrases must not be used. Instead, the author is considered to have published a later (junior) homonym that must be ascribed solely to him. Retention of such a name can be effected only by conservation.

7.9 Names of subordinate taxa containing the type of the name of the immediately higher taxon

The subordinate taxon containing the type of the name of the higher taxon to which it is subordinate is called (by the Zoological Code) the *nominate* subordinate taxon and under all three Codes its name must repeat the name (except, where appropriate, for suffix) or epithet of the higher taxon. Considered as subordinate taxa for the purposes of this rule are all taxa of categories the names of which begin with the prefixes *sub-* and *infra-* and all other categories below ordinal rank except Superfamily, Family, Genus and Species.[28]

This rule is responsible for names such as *Culicidae* subfamily *Culicinae*, *Bacillus* subgenus *Bacillus* and *Salix repens* subsp. *repens* which sometimes puzzle the reader by their repetition. They indicate merely that the subordinate taxon concerned includes the type of the name of the higher taxon. For example, the genus *Culex* is the type genus of the family name *Culicidae*; the family *Culicidae* is divided into a number of subfamilies, of which the one within which the genus *Culex* is classified must be called subfamily *Culicinae*. The type species of the generic name *Bacillus* is *B. subtilis*. If the genus *Bacillus* is divided into subgenera, then the subgenus within which this species is classified must be called subgenus *Bacillus*.

One point to be noted about the names of nominate subordinate taxa is that under the Botanical Code they are cited without authorities. This is because a taxon is considered nomenclaturally to be the sum of all its subordinate taxa. Therefore the publication of the name of a subordinate taxon which does not include the type of the next higher taxon is considered automatically to establish the name of another subordinate taxon which is the nominate subordinate taxon. Thus the publication by A. and G. Camus of the subspecific name *Salix repens* L. subsp. *argentea* (Sm.) A. et G. Camus is considered automatically to have established the name *Salix repens* L. subsp. *repens* for the nominate subspecies. When we write simply *Salix repens* L. we imply we are referring to the *whole* of the species *Salix repens*, i.e. to subsp. *argentea*, subsp. *repens* and all the other subspecies that there may be. If we wish to refer to only that part of *Salix repens*. L which is considered subspecifically identical with the type specimen of the specific name, then we must write *Salix repens* L. subsp. *repens*. Under the Bacteriological Code, the author of the name of the species is to be cited as the author of such an automatically established subspecific name.

At one time it was common to denote nominate subordinate taxa by the use of the prefix *Eu-* or by such words as *typicus, originalis* or *genuinus*, often quoted with their publishing authorities. Such names will be found in the older literature, but these practices are no longer permitted by the Codes.

8

Special Cases

8.1 Certain plant families

Eight plant families are exceptions in that each has two alternative names, both of which are correct under the Botanical Code. One is a standard name, ending in -*aceae*; the other is an exception, sanctioned by long usage. These and their alternatives are the following: *Palmae* (*Arecaceae*); *Gramineae* (*Poaceae*); *Cruciferae* (*Brassicaceae*); *Leguminosae* (*Fabaceae*); *Guttiferae* (*Clusiaceae*); *Umbelliferae* (*Apiaceae*); *Labiatae* (*Lamiaceae*); and *Compositae* (*Asteraceae*).

In addition, when the *Papilionaceae* are recognized as a family distinct from the rest of the *Leguminosae*, this name is conserved over *Leguminosae*; *Fabaceae* is then the alternative name for *Papilionaceae*.

8.2 Fossil plants

The names of fossil plants under the Botanical Code are subject to the same rules as those of living plants. However, because fossil plants usually occur as detached organs or fragments of organs, the names usually refer to parts rather than to complete organisms. The Code therefore recognizes a special category—the *form-genus*—under which species may be recognized and given names. A form-genus may be unassignable to a family, but may be referable to a taxon of higher rank. It may include fossils superficially similar but in fact representing widely different taxa of family and higher rank. For example, the form-genus *Stigmaria* is used for the rhizophoric organs of various families of fossil arborescent lycopods. Unless the organ is found actually attached to the stem (which is but rarely), it is impossible to tell to which trunk genus a *Stigmaria* may correspond. The Code permits the use of form-genera as means of reference to such isolated and unassignable parts.

8.3 Fungi

The names of *Fungi* are governed by the Botanical Code, and are subject to the same rules as are the names of plants. Provision is made, however, for the separate naming of perfect and imperfect states of *Ascomycetes* and

Basidiomycetes. Many of these fungi are well known in the imperfect (asexual, conidial) state but are unknown (or rarely seen) in the perfect (sexual, ascal or basidial) state. It is therefore convenient to give separate names to the imperfect and perfect states and this is permitted by the Code. However, while the name for the imperfect state may be used as necessary to refer to that state alone, the name of the perfect state must be used for the fungus when both states are being referred to. Generic and specific names of imperfect states may not be used to refer to the perfect states, nor do they compete with names given to the latter for purposes of priority.

The Botanical Code also makes provision for the recognition of special forms (*formae speciales*) within parasitic species to distinguish physiologic variants characterized by their adaptation to different hosts. However, it does not regulate the nomenclature of such special forms.

8.4 Lichens

The names of lichens (*Lichenes*) are governed by the Botanical Code. For nomenclatural purposes, the names given to lichens are considered to apply to their fungal components. They are thus also applicable to the lichen fungi in their non-lichenized condition, should they ever so occur, and compete with ordinary fungal names for the purposes of priority. Names given to lichenized fungal taxa which already have a legitimate lichen name are therefore, in general, nomenclaturally superfluous and illegitimate. Lichen algae must bear independent names.

8.5 Cultivated plants

The names of plants in cultivation are governed by the Botanical Code when they refer to taxa, such as species, which belong to categories covered by the Code (see Table 1, p. 3). In this, they do not differ from the names of plants occurring in the wild. However, cultivated plants do not have a natural population structure; they occur as artificial populations maintained and propagated by man. For this reason, the botanical hierarchy of infraspecific categories is hardly applicable to cultivated plants. It is largely replaced by a system based on the taxonomic category *cultivar*.

Cultivar is the internationally recognized term for the category of distinct cultivated sorts, those the grower talks about when he orders from his seedsman or nurseryman and discusses and sells his crop. They are usually called 'varieties' in trade catalogues and in Britain they are subject to the provisions of the Plant Varieties and Seeds Act. The term cultivar is much to be preferred to the English word variety, as the latter could easily be confused with the quite different botanical category *varietas*, which is also called variety in English. The names of cultivars are governed by a separate Code, the International Code of Nomenclature of Cultivated Plants.

A cultivar is any assemblage of cultivated plants which is clearly distinguished by any characters and which retains its distinguishing characters when reproduced, sexually or asexually. A cultivar may be a clone, a group of indistinguishable clones, a line or lines of selfed or inbred individuals, a series of cross-fertilized individuals, or an assemblage of individuals reconstituted on each occasion by crossing, such as an f_1 hybrid. The mode of origin of a cultivar is immaterial. Some have originated in the wild, but the majority have arisen in cultivation, either spontaneously or as induced sports or artificial hybrids. The important point is that they are *maintained* by cultivation.

Cultivar names are written with capital inital letters. They are preceded by the abbreviation cv. (cultivar) or placed in single quotation marks. They may be used with generic, specific or common names if these are unambiguous, and with the names of hybrids (see §8.6, p. 44). The following are typical: *Citrullus lanatus* cv. Sugar Baby or *Citrullus* cv. Sugar Baby or water melon cv. Sugar Baby; *Cucurbita pepo* 'Table Queen' or *Cucurbita* 'Table Queen' or pumpkin 'Table Queen'.

Cultivar names must be fancy names, in modern languages, not Latin names such as 'Totus Albus'. Prior to 1 January 1959, such Latin names could be given, so many are still in use, but they may not be given to new cultivars. The only exceptions are names of botanical taxa reduced to cultivar rank. Cultivar names must be published by the distribution to the public of printed or similarly duplicated matter dated at least to year. A description, or a reference to a description, in any language, is also required. New cultivar names must not be the same as the botanical or common name of a genus or the common name of a species if confusion might be caused. For example, names such as poplar 'Eucalyptus' or camellia 'Rose' are not permitted, but a name like carnation 'Heather Pink' is admissible as the likelihood of confusion is small. A cultivar name that is established by legal process, such as entry in a statutory register, is considered by the Cultivated Code to be in accordance with the articles of the Code irrespective of any other provisions of the Code, unless the cultivar of which it purports to be the name neither exists nor did exist.

Each cultivar has one correct name, the name by which it is internationally known. Cultivar names are subject to the ordinary operation of priority and synonymy; mode of origin is irrelevant as a consideration in deciding whether two or more cultivars should be considered synonymous. Exceptions may be made to the operation of priority if general usage so demands. A cultivar name may also have one or more legitimate synonyms. A *commercial synonym* may be used instead of the correct name in a country where the latter is commercially unacceptable. Cultivar names may also be transliterated or translated, although the Code does not recommend this; such translations or transliterations are regarded as different forms of the original names, and their dates are those of the originals. The date of a cultivar name is that of its valid publication, or of its registration if that is earlier and before 1 January 1959.

Only one cultivar may bear a given name in any one cultivar class. By *cultivar class* is meant the taxon within which the use of the same cultivar names for two distinct cultivars would create confusion. A cultivar class may correspond to a genus, a species, a crop type or a group of cultivars; the limits of cultivar classes are fixed by the appropriate registration authority or in the absence of one, by the International Commission for the Nomenclature of Cultivated Plants. Thus we could not name a primrose cultivar *Primula vulgaris* 'Lilac Queen' as there is already a *Primula malacoides* 'Lilac Queen' and either could be referred to as simply *Primula* 'Lilac Queen'. On the other hand, we could have a cabbage 'Favourite' as well as a cauliflower 'Favourite', as these are two distinct crops and no confusion would result, even though both are members of the same genus, *Brassica*.

Cultivar names remain unchanged when botanical names change. For example, cornflower 'Blue Diadem' retains its name, regardless of whether the species is called, botanically, *Centaurea cyanus*, or *Cyanus segetum*.

Taxa below the rank of cultivar are not recognized by the Cultivated Code. Any selection of a cultivar showing sufficient differences from the parent cultivar to render it worthy of a distinct name must be regarded as a distinct cultivar; the practice of designating such a selection by the term 'strain' or its equivalent is not recognized by the Code.

The Cultivated Code, like the Bacteriological, Botanical and Zoological Codes, does not have any legal status in national or international law, and the observance of the provisions of its Articles is purely voluntary. However, there exist also certain statutory regulations concerning the names of cultivars of those groups of plants in which plant breeders' rights have been recognized. The basis for such regulations are the Guidelines for Variety Denominations approved by the Council of the International Union for the Protection of New Varieties of Plants (UPOV). The responsibility for the implementation of these guidelines lies with the government of each of the member states of UPOV by way of the appropriate statutory body established for the purpose. In Great Britain it lies with the Controller of Plant Variety Rights through the Plant Variety Rights Office. In the USA, the Plant Variety Protection Office has similar functions. The current UPOV Guidelines are published in the *Plant Varieties and Seeds Gazette*, **109**: 1–3 (Feb. 1974). The term 'variety denomination' as used here is the equivalent of the term 'cultivar name' of the Cultivated Code. Basically, the UPOV Guidelines resemble the Articles of the Cultivated Code regulating the formation of cultivar names, but are more detailed and specific in certain respects. Cultivar names given to new cultivars (varieties), ideally should be such as to conform with the requirements of both the UPOV Guidelines and the Cultivated Code.

An important feature of the Cultivated Code is its provision for the *registration* of cultivar names with a recognized registration authority which undertakes to keep a list (register) of cultivar names for the group concerned. Registration authorities may act either nationally or internationally. For example, the Royal Horticultural Society acts as the international registration authority for the names of cultivars of the genus *Lilium*. Where plant breeders' rights have been legally recognized in a group of cultivated plants, the appropriate statutory body may perform the functions of a registration authority. In Great Britain, for example, the Plant Variety Rights Office has taken over the functions of British national registration authority for field crops, formerly carried out by the National Institute of Agricultural Botany, although the latter still acts as a testing and verification authority for new varieties (i.e. cultivars) submitted for the granting of breeders' rights.

The Cultivated Code may be modified only by action of the International Commission for the Nomenclature of Cultivated Plants of the International Union of Biological Sciences; proposals for amendment of the Code should be sent to the Secretary of the Commission.

8.6 Hybrids

The Zoological Code expressly excludes names given to hybrids from its province. It does specify, however, that a species-group name, if otherwise available, remains available even if it is found that the original description relates to an animal or animals later found to be hybrid. Such a name must not, however, be used to denote either of the parents of the hybrid. The Bacteriological Code likewise makes no provision for the naming of hybrids. On the other hand, the Botanical and Cultivated Codes are concerned with the nomenclature of hybrids in the plant kingdom.

Under the Botanical and Cultivated Codes, hybrids produced by sexual crossing are indicated by the use of the multiplication sign (×). They may be designated either by *names* or by *formulae*.

Hybrids between species of the same or different genera may be designated by a formula which is simply the names of the parent species connected by the × sign, e.g. *Digitalis purpurea* × *D. lutea, Cupressus macrocarpa* × *Chamaecyparis nootkatensis*. On the other hand, hybrids between species of the same genus may be given distinct names of their own, which resemble the names of species but have an × sign between the generic name and the collective (hybrid specific) epithet, e.g. *Verbascum* × *schiedeanum*, otherwise *V. lychnitis* × *V. nigrum*.

When hybrids between species of two different genera are involved, the two correct names of the currently accepted genera are combined either into a formula, e.g. *Asplenium* × *Phyllitis*, or into a condensed formula, preceded by the × sign, e.g. × *Aspleniophyllitis*. The same applies to names of subgenera, sections and other subgeneric supraspecific taxa. Condensed formulae may also be used to designate trigeneric hybrids, e.g. × *Dialaeliocattleya* (*Diacrium* × *Laelia* × *Cattleya*), but these can be given names, identical in status with condensed formulae, formed from the names of persons, preceded by ×, and ending in *-ara*, e.g. × *Sanderara* (*Brassia* × *Cochlioda* × *Odontoglossum*). Such names must be used instead of condensed formulae for hybrids derived from four or more genera, e.g. × *Carrara* (*Rhynchostylis* × *Asocentrum* × *Euantha* × *Vanda*).

Such condensed formulae (including *-ara* names) need be accompanied only by a statement of parentage, in lieu of a description, as a condition of valid publication. They are used in combination with an appropriate collective epithet to refer to hybrids between species of two or more genera, e.g. × *Arachnopsis rosea* (*Phalaenopsis schillerana* × *Arachnis maingayi*).

The binary names given to hybrids are subject to the same rules as are the names of species, and authorities are cited for them in the usual way. A change in status from hybrid to species, or *vice versa*, does not entail a change in the authority citation.

Names of graft hybrids (or chimaeras) are regulated by the Cultivated Code. Graft hybrids are named in exactly the same way as are sexual hybrids, except that the plus sign (+) is used in place of the × denoting a sexual cross, e.g. *Syringa* + *correlata*, + *Crataegomespilus dardarii* (or *Crataegus monogyna* + *Mespilus germanica*).

Under the Cultivated Code, assemblages of similar cultivars may be designated as *groups*. Group names may be placed in parentheses before the cultivar name, e.g. *Lolium perenne* (Early group) 'Devon Eaver'. No provisions are made by the Code for the naming of taxa of any other categories.

The Cultivated Code also regulates the formation and use of collective names which refer to groups of cultivars of hybrid origin. Such collective names consist of a generic name or condensed formula followed by a collective epithet which must be a word or phrase of not more than three words of a modern language. A phrase used as a collective epithet may contain a word, e.g. Hybrid, Hybrids, grex (abbreviated g.) indicative of the collective nature of the taxon to which it refers, e.g. *Lilium* Bellingham Hybrids (*L. humboldtii* × *L. pardalinum*). When used with cultivar names, such collective epithets should normally be placed in parentheses, e.g. *Lilium* (Bellingham Hybrids) 'Shuksan', *Cattleya* (Fabia grex) 'Prince of Wales'. Where the status of the collective epithet is clear, both the parentheses and the word grex (or its abbreviation) may be omitted, e.g. *Cattleya* Fabia 'Prince of Wales'.

8.7 Cultivated orchids

The Handbook on Orchid Nomenclature and Registration extends, amplifies and clarifies the provisions of the Botanical and Cultivated Codes to cover problems especially affecting the naming of cultivated orchids. It makes especial provision for the formation, use and regulation of grex names, which are used in the naming of *artificial hybrids* of cultivated orchids. The same *grex name* is applied to all the progeny raised from any, each and every crossing of any two parent plants belonging to different taxa that bear the same pair of specific, hybrid-specific or grex names. In cultivated orchids the cultivar class (see §8.5, p. 43) is a grex, a natural hybrid or a species; the 'grex class' is a genus or a hybrid genus. The international registration authority for cultivated orchids is the Royal Horticultural Society, London, and it is grex names, not cultivar names, that are registered. The Orchid Handbook also includes a list of recommended names (of species and genera) that are being used by the registration authority in current and future grex registrations. Some of these differ from the names currently considered botanically correct for the taxa concerned.

8.8 Infrasubspecific taxa of bacteria

The names given to infrasubspecific taxa of *Bacteria* are not governed by the Bacteriological Code. Such taxa are given vernacular names, or, in some cases, names in Latin form, or are designated by numerals or letters. The Code, however, strongly recommends bacteriologists to use each of the different infrasubspecific categories in a generally accepted sense, and gives guidance as to the use of categories such as strain, type (in the non-nomenclatural sense), group, phase, *forma* (or *forma specialis*) and state (or stage). The use of the suffix '-var' (or '-form') to replace '-type' is recommended to avoid confusion with the strict nomenclatural use of the latter. Thus it is recommended that, for example, morphovar, phagovar and serovar should be used instead of, respectively, morphotype, phagotype and serotype.

8.9 Domesticated animals

The names of domesticated animals are governed by the Zoological Code when they refer to taxa, such as species, which belong to categories covered by the Code (see Table 1, p. 3). In this respect, the Zoological Code resembles the Botanical Code, although it has recently been suggested that the names of domesticates should be excluded from its coverage. On the other hand, there is nothing in zoology that corresponds to the International Code of Nomenclature of Cultivated Plants—i.e. there is no internationally recognized code for the breed names of domesticated animals. For livestock, there is an unofficial standard which many people in the English-speaking world use. It is the *Dictionary of Livestock Breeds*, by J. L. Mason, published by the Commonwealth Agricultural Bureaux, Farnham Royal, Bucks., England. For kinds of livestock within breeds, the use of vernacular names should be avoided wherever possible. The British Society of Animal Production has produced a list of recommended nomenclature for kinds and classes of livestock of different ages, and this should be followed.

8.10 Viruses

A universal system of nomenclature for viruses is still in the process of establishment. Virology developed from pathology and viruses became designated by names formed from those of the diseases (or symptoms) they produced. In consequence, viral nomenclature developed piecemeal, varying in practice according to the kind of organism infected and the kind of symptons observed, and became burdened by a large synonymy.

In an attempt to rectify a situation becoming ever more chaotic, an International Committee on Nomenclature of Viruses was set up in 1966, and the first report—P. Wildy, *Classification and Nomenclature of Viruses*

(Monographs in Virology, vol. 5) was published in 1971. This was intended as a first step in the establishment of a system of viral nomenclature that could be universally useful and acceptable.

Since the publication of this report, the International Committee on Nomenclature of Viruses has become the International Committee on Taxonomy of Viruses (ICTV) with amongst its functions the approval for universal use of formally-named viral taxa of generic rank and above. Official approval for new names follows recommendation by one or more of the subcommittees of the Executive Committee of the ICTV (sub-committees on Bacterial, Invertebrate, Plant, and Vertebrate Viruses and a Coordination Sub-Committee have been established), consideration by the Executive Committee and finally, acceptance by the International Committee on Taxonomy of Viruses itself. Only after this final approval does a name become official.

The 18 provisional rules of viral nomenclature set out in Wildy's report have been modified in the light of experience and at the 3rd International Congress for Virology at Madrid in September 1975 a revised set of 16 Rules was accepted and is now current. The official approval of names for viral taxa is subject to their legitimacy in terms of these rules. Viral nomenclature is to be independent of the Bacteriological Code, international and universally applied to all viruses. The adoption of a latinized nomenclature is recommended as far as possible for names of families and genera, but not for the names of species. The principle of priority is not to be applied. Names of families are to end in the suffix *-viridae*, those of genera, in the suffix *-virus*. For details of the Rules, see F. Fenner, *The Classification and Nomenclature of Viruses* (*J. Gen. Virol.*, **31**: 463–470, 1976), which summarizes the results of the Madrid meetings of the ICTV. A summary of the results to date of the work of the ICTV and its committees is due to be published early in 1977 as a separate number of *Intervirology*, the official journal of the Section on Virology of the International Association of Microbiological Societies.

Fenner's 1976 article is *de facto* almost an international code of viral nomenclature and as such it is recommended to the attention of all who deal with viral names. For the names of viral species, standardized lists of viral names (where they exist) should be followed, particularly if they give a critical synonymy, otherwise general practice should be followed and new viral names should not be published until adequate tests have been performed to ensure that a new viral manifestation is not being caused by an already named virus.[29]

8.11 Exceptions

This Handbook has attempted to give an outline of biological nomenclature, but it is only an outline. It must not be used as a substitute for the Codes, which must be consulted on all matters of detail. It is also an idealized outline. Numerous examples of nomenclatural practice will be found in biological literature which are not in accordance with the requirements of the appropriate Code. There are several reasons for this. Some are historical. The Codes have evolved and their requirements have changed. Most of their provisions are retroactive, so past exceptions to currently required practice are inevitable. Moreover, at times different Codes have been followed simultaneously by different groups of workers. For example, during the years 1892–1930, two conflicting Codes of nomenclature were current among botanists, one of which, the so-called American Code, did permit the use of tautonyms. A truly cosmopolitan set of zoological rules was not adopted until 1898, previous to which little unanimity existed. There have also been various specialist Codes in the zoological field, such as the Ornithological and Entomological Codes. Such restricted Codes are useful as extensions and amplifications to the three main Codes, provided that they do not run counter to the latter in any important respect. Unfortunately, this condition has not always been observed.

Exceptions also arise from the failure of biologists to follow the Codes. This may be a result of ignorance, carelessness, lack of concern with nomenclature, or of such strong disagreement with some provision of a Code that it is deliberately ignored or counteracted. Such practices are to be deplored, for they inevitably cause only confusion and difficulty for later workers. However, should systematists adopt some system of classification other than the hierarchical (or 'Linnaean') systems that the Codes of Nomenclature are designed to serve, then they not only may but ought to adopt systems of nomenclature (or numericlature) other than those laid down by the Codes. Such systems are outside the scope of this book, as is the nomenclature of diseases and other pathological manifestations.

There are also a few cases in which the Codes are genuinely ambiguous and their interpretation equivocal. For this and the above reasons, a word of warning must be added in conclusion. A proper understanding of nomenclatural practice will add immensely to the amount of information users of systematic literature can extract from the formal nomenclatural sections. But caution is necessary and in case of doubt an appropriate authority should always be consulted.[30]

Notes to the Text

These notes are referred to in the text by the corresponding small 'superior' figures ([1], [2], ...).

1. Systematics is sometimes referred to as *taxonomy*, but taxonomy is, strictly speaking, the study of the principles and practice of classification. As such, it is only a part of systematics. However, in practice the terms systematics and taxonomy are commonly used as if they were synonymous.
2. The word 'classification' is also used to denote the result of this process, i.e. a system of classification, as in the phrase 'Hutchinson's classification of the flowering plants'.
3. Such a system can also be referred to as hierarchal.
4. A taxon is defined as a taxonomic group of any rank; other equivalent phrases are taxonomic grouping, taxonomic entity, and taxonomic unit.
5. The word 'taxon' is sometimes used in the sense of 'a taxonomic *category* of any rank', but this usage is best avoided, to obviate confusion, although the intended sense is usually clear from the context.
6. For some groups of animals, e.g. birds and domesticated mammals, vernacular names are frequently used in scientific and technical literature, although it is usual for the scientific name to be given as well when a species is first mentioned in a given paper. In such groups, the concepts referred to by traditional vernacular names accord fairly closely to the species recognized by zoologists. In other groups, such as most of the invertebrates and the plants, most species have no common name and where they do exist, common names tend to correspond to genera and higher taxa rather than to species. There are systems of standardized common names for various groups which avoid many of the defects of ordinary vernacular names, but the majority of such names have been especially coined (often by translation of the scientific name) and are not common names used by laymen in everyday speech.
7. This was formerly the International Committee on Bacterial Nomenclature.
8. By Latin is meant here biological Latin, which is as distinct a language from classical Latin as is modern English from Chaucerian. It should be regarded as a modern, autonomous technical language, derived from Renaissance Latin, much enriched by a host of words, derived from Greek and other sources, unknown to classical Latin, using many classical Latin words in new, specialized senses, simplified in grammar, and using letters—j, k, and w—not employed in classical Latin.
9. There are 8 non-standard exceptions, sanctioned by long usage, which are permitted as correct alternatives to names ending in *-aceae*, e.g. *Compositae* for *Asteraceae*; *Gramineae* for *Poaceae* (see p. 40).
10. Or, if necessary, of the immediately higher taxon in the case of a taxon below the rank of subgenus.

11. In the Botanical and Bacteriological Codes, the complete binary name is referred to as the *specific name,* and the second term as the *specific epithet.* In the Zoological Code, the complete binary name is referred to as the *binomen,* and the second term as the *specific name.* This difference in usage should be noted, as it is a possible source of confusion.

12. The Botanical and Bacteriological Codes agree in recommending that in the latinization of non-classical names of persons ending in a consonant the genitive is formed by adding *-ii,* except in the case of those ending in *-er* when *-i* is added. The Zoological Code recommends the addition of *-i* in all cases. The same recommendations apply to the feminine and plural forms of the endings, except that under the Bacteriological Code, *-ae* instead of *-iae* is added to feminine names ending in a consonant. The Botanical Code alone of the three rules that the wrong use of these endings be treated as orthographic errors which must be corrected.

13. The Bacteriological and Botanical Codes recommend that the second term should be written with a small initial letter, though they permit a capital to be used if it is derived from the name of a person (both Codes) or (Botanical Code only) is a vernacular (non-Latin) name or a former generic name. The Zoological Code makes a small initial letter obligatory in all cases.

14. The Zoological Code does, however, give rules that determine whether subspecific or infrasubspecific status should be accorded to the name of any given taxon of infraspecific rank.

15. In the Botanical and Bacteriological Codes, the complete ternary name is referred to as the *subspecific name,* and the third term as the *subspecific epithet.* In the Zoological Code, the complete ternary name is referred to as the *trinomen,* and the third term as the *subspecific name.* This difference in usage parallels that already noted with respect to the names of the species.

16. In sexually reproducing organisms, species, unlike taxa of other ranks, are *non-arbitrary as to both inclusion and exclusion.* That is, within species there is internal continuity in variation and breeding pattern, between species there is external discontinuity in these features. Taxa above the rank of species are arbitrary as to inclusion (with internal discontinuities); taxa below the rank of species are arbitrary as to exclusion (with external continuities).

17. Care must be taken not to confuse the term 'valid' as defined by the Zoological Code (which is the equivalent of the term 'correct' of the Botanical and Bacteriological Codes) with 'valid' as defined by the Bacteriological Code, which is the equivalent of 'available' under the Zoological Code and of 'validly published' under the Bacteriological and Botanical Codes.

18. Under the Botanical Code, this is so for taxa of the rank of subgenus and below.

19. The term illegitimate is also employed by the Bacteriological and Botanical Codes in a rather wider sense, to denote any name not in accordance with the Rules of the Code.

20. Or in a manner so similar as to be considered identical under the provisions of the appropriate Code.

21. This is so even if the oldest homonym is itself, for some other reason, illegitimate, and therefore itself also excluded from use.

22. After 1 Jan. 1980, names of *Bacteria* validly published under the current edition of the Bacteriological Code are not to be regarded as homonyms of names of *Bacteria* published under previous editions of the Code.

23. An example of the working of the superfluous names and tautonyms rules in botanical nomenclature is afforded by the species for which the correct name, when it is included in the genus *Radiola*, is *Radiola linoides* (1788). This was originally called *Linum radiola* (1753) but the epithet '*radiola*' cannot be used in combination with *Radiola* as a tautonym would result. *Linum multiflorum* (1778) is the next oldest name for the taxon but it happens to be a superfluous name for *Linum radiola*. It is therefore illegitimate and must not be taken into consideration for the purpose of priority. The epithet '*multiflorum*' dating from 1778 cannot in consequence be used in *Radiola* to displace '*linoides*' although the latter dates from 1788. *Radiola linoides* is therefore the correct name for the taxon.

24. The setting-up of a committee to report on the desirability and practicality or otherwise of both conservation and rejection of the names of species has been authorized by the 12th International Botanical Congress (Leningrad, July 1975). This committee is to report to the 13th International Botanical Congress.

25. The only exceptions are names to be rejected as widely and persistently misapplied (see §5.15), rejection of which is possible no matter what the rank of the taxa concerned.

26. Under the Zoological Code, such variants are considered to be forms of the same name only if the change in orthography is not demonstrably intentional.

27. If, after 1 Jan. 1980, a name or epithet published prior to that date but not included in an Approved List is re-used by an author for the same taxon, he may, if he wishes, indicate the name of the original author of the revived name by placing, in parentheses after the name of the taxon, *ex* followed by the name of the original author and, after the parentheses, the abbreviation 'nom. rev.' (nomen revictum), e.g.

Bacillus palustris (*ex* Sickler & Shaw 1934) nom. rev.

After 1 Jan. 1980, under the Bacteriological Code, *ex* may be used in subsequent citation of such names in the following manner:

Bacillus palustris (*ex* Sickler & Shaw 1934) Brown 1982.

This shows that Brown, in 1982, when publishing his name, cited the names of the original authors in the above manner and stated that he was reviving their name for what in his opinion was the same taxon.

28. The Botanical Code alone provides for certain exceptions to which this rule is not to be applied.

29. The use of the standard cryptogram (see glossary) which has been developed for summarizing viral properties is also to be recommended.

30. Students of botany and zoology wishing to make use of taxonomic literature in conjunction with this Handbook should consult the following bibliography, published by the Systematics Association: Kerrich, G. J., Hawksworth, D. L. and Sims, R. G. (eds.), *Key Works to the Fauna and Flora of the British Isles and North-West Europe*, London, 1977 (*in the press*). Fossils are unfortunately not included.

Bibliography

This bibliography lists the official publications on nomenclature and the periodicals associated with them; additional sources used in the compilation of this handbook are also given.

Bacteriology

International Code of Nomenclature of Bacteria (1975). American Society for Microbiology, Washington, DC. (Available from the American Society for Microbiology, Publications Office, 1913 I Street, N.W., Washington, DC. 20006, USA.)
International Journal of Systematic Bacteriology. Ames, Iowa.

Botany

International Code of Botanical Nomenclature (1972). International Association for Plant Taxonomy, Utrecht. (Available from the International Bureau for Plant Taxonomy and Nomenclature, Tweede Transitorium, Uithof, Utrecht, Netherlands.)
International Code of Nomenclature of Cultivated Plants (1969). *Regnum Vegetabile*, **64**, 1–32. (Available from the International Bureau for Plant Taxonomy and Nomenclature, Tweede Transitorium, Uithof, Utrecht, Netherlands.)
Handbook on Orchid Nomenclature and Registration (1969). International Orchid Commission, Cambridge, Mass. (Available from the International Orchid Commission on Classification Nomenclature and Registration, Botanical Museum of Harvard University, Cambridge, Mass. 02138, USA.)
Regnum Vegetabile. Utrecht.
Taxon. Utrecht.

Virology

FENNER, F. (1976). The Classification and Nomenclature of Viruses. *J. Gen. Virol.* **31**: 463–470 (1976). (Available in reprint form from Dr. J. Maurin, Service des Virus, Institut Pasteur, Paris, France.)
Intervirology. Basle.

Zoology

International Code of Zoological Nomenclature (1964). International Trust for Zoological Nomenclature, London. (Available from the International Trust for Zoological Nomenclature, c/o British Museum (Natural History), Cromwell Road, London SW7 5BD, England.)
Official Index of Rejected and Invalid Family-Group Names (1958). First Instalment, ed. F. HEMMING (1966). Second Instalment, ed. W. E. CHINA. International Trust for Zoological Nomenclature, London.

Official Index of Rejected and Invalid Generic Names (1958). First Instalment, ed.
F. HEMMING. (1966). Second Instalment, ed. W. E. CHINA. International Trust for
Zoological Nomenclature, London.
Official Index of Rejected and Invalid Specific Names (1958). First Instalment, ed.
F. HEMMING (1966). Second Instalment, ed. W. E. CHINA. International Trust for
Zoological Nomenclature, London.
Official Index of Rejected and Invalid Works in Zoological Nomenclature (1958). Ed.
F. HEMMING. International Trust for Zoological Nomenclature, London.
Official List of Family-Group Names in Zoology (1958). First Instalment, ed.
F. HEMMING (1966). Second instalment, ed. W. E. CHINA. International Trust for
Zoological Nomenclature, London.
Official List of Generic Names in Zoology (1958). First Instalment, ed. F. HEMMING
(1966). Second Instalment, ed. W. E. CHINA. International Trust for Zoological
Nomenclature, London.
Official List of Specific Names in Zoology (1958). First Instalment, ed. F. HEMMING
(1966). Second Instalment, ed. W. E. CHINA International Trust for Zoological
Nomenclature, London.
Official List of Works Approved as Available for Zoological Nomenclature (1958). Ed.
F. HEMMING. International Trust for Zoological Nomenclature, London.
The Bulletin of Zoological Nomenclature. London.

Sources

AINSWORTH, G. C. (1968). Names. *Rev. Med. Vet. Mycol.* **6** (8), 379–85.
BLACKWELDER, R. E. (1967). *Taxonomy.* John Wiley, New York.
HAWKSWORTH, D. L. (1974). *Mycologist's Handbook.* Commonwealth Mycological
Institute, Kew.
HEYWOOD, V. H. (1976). *Plant Taxonomy.* 2nd edn. Edward Arnold, London.
JEFFREY, C. (1968). *An Introduction to Plant Taxonomy.* Churchill, London.
MCCLINTOCK, D. (1969). *A Guide to the Naming of Plants.* Heather Society, Horley.
MCVAUGH, R., ROSS, R., and STAFLEU, F. A. (1968). An Annotated Glossary of Botanical
Nomenclature. *Regnum Vegetabile* **56**, 1–31.
MELNICK, J. L. (1975). Taxonomy of Viruses, 1975. *Progr. Med. Virol.*, **19**: 353–8.
ROSS, R. (1962). Nomenclature and Taxonomy. In AINSWORTH, G. C., and SNEATH,
P. H. A. (eds.), *Microbial Classification.* University Press, Cambridge.
SAVORY, T. (1962). *Naming the Living World.* English Universities Press, London.
STEARN, W. T. (1966). *Botanical Latin.* Thomas Nelson, London.
WILDY, P. (1971). Classification and Nomenclature of Viruses. *Monographs in
Virology* (ed. J. C. Melnick), **5**. S. Karger, AG., Basle (also available through
Academic Press, London).
ZABINKOVA, N., and KIRPICHNIKOV, M. (1957). *Vademecum Methodi Systematis
Plantarum Vascularum. Fasc. II. Lexicon Latino-Rossicum pro Botanica.* Academy
of Sciences of the USSR, Moscow–Leningrad.

Glossary/Index

absolute synonym: *see* synonym, homotypic.

actinomycetes, names of, 5.

admissible: (Bot.) of a name, of a form that could permit it to be validly published; of the use of a name or epithet, in accordance with the provisons of the code.

aff., affinis: lat., akin to.

al., alii, aliorum: lat., others, of others.

alphabet, names, of, 8.

ambiguous name, 25: *see* nom. ambig.

analysis: (Bot.) a figure or figures, illustrative of a taxon, and showing the details necessary for identification.

animals, names of, 5.

ap.: *see* apud.

application, 18: the use of a name to denote a taxon.

approved: 1. (Zoo.) given approval by the International Commission on Zoological Nomenclature for use in nomenclature.

    2. (Bact.) of a name, validly published before 1 January 1980 and given approval by the International Committee on Systematic Bacteriology for use in nomenclature; of a list, made up of such names.

apud: lat., with, in the work of; *see* in.

article, 5: (Bot., Cult., Zoo.) a numbered section of the Code, consisting of a rule or rules, which are mandatory, and sometimes also of examples, which are explanatory, and/or supplementary recommendations (*q.v.*)

auct., auctorum: lat., of authors.

auct. non., auctorum non: lat., of authors, not (of); used to indicate a misapplied (*q.v.*) name.

author, of a name, 32: the person who first published a name in such a way as to satisfy the criteria of availability (Zoo.) or valid publication (Bact., Bot.)

authority, 32: the name of the author (*q.v.*) of a name, cited as such, after the name, usually in abbreviated form.

    authority, citation of, 32.

citation, authorities, of, 32.

citation, dates of publication, of, 35.

citation, double, 33.

citation, homonyms, of, 35.

citation, misidentifications, of, 37.

citation, parentheses, use of in, 33.

citation, punctuation, use of in, 36.

citation, square brackets, use of in, 34.

citation, subsequent usages, of, 36.

citation, synonyms, of, 38.

classification, 2: 1. The process of establishing and delimiting taxa (*q.v.*)
    2. A system (*q.v.*) so produced.

cl., clone: 1. (Bot., Cult.) a group of individuals formed by the vegetative (natural or artificial) or apomictic reproduction of a single original parent.
    2. (Bact.) a population of bacterial cells derived from a single parent cell.

clone: *see* cl.

code, nomenclature, of, bacteriological, 5.

code, nomenclature, of, botanical, 5.

code, nomenclature, of, cultivated plants, of, 42.

code, nomenclature, of, zoological, 5.

codes, nomenclature, of, 5.

codes, nomenclature, of, aims of, 16.

codes, nomenclature, of, exceptions to, 49.

codes, nomenclature, of, modification of, 6.

codes, nomenclature, of, specialist, 49.

collective: (Bot. Cult., Orch.) of names or ephithets, used to denote hybrids or groups of hybrids; (Zoo.) *see* collective group.

collective group: (Zoo.) an assemblage of identifiable species of which the generic positions are uncertain; accorded a collective name, treated as a generic name in the meaning of the Code, but which requires no type species: analogous to a form-genus of Bot. (*q.v.*).

combination, 10: the name of a taxon of below generic rank (Bot.) or of specific or subspecific rank (Bact., Zoo.), consisting of the name of the genus followed by one or more words peculiar to the taxon; *see* epithet (*q.v.*).

comb. nov., combinatio nova: lat., new combination; a combination (*q.v.*) made available (Zoo.) or validly published (Bact., Bot.) for the first time ,and based on a previously available or validly published combination (basionym, *q.v.*), from which the word peculiar to the taxon (epithet, *q.v.*) is transferred; used in citation to indicate a change in the position and/or rank of a nominal taxon (*q.v.*).

commercial synonym: *see* synonym, commercial.

common names, 5.

condensed formula: *see* formula, condensed.

confused name, 25: *see* nom. conf.

conservation, 24: procedure whereby the use of a name which would in the absence of such procedure contravene the provisions of a Code is made possible.

conserved name, 25: *see* nom. cons.

consortium: (Bact.) an aggregate or association of two or more organisms; names given to consortia are not regulated by the Bacteriological Code, and have no standing in bacteriological nomenclature.

coordinate, 11: (Zoo.) of equal nomenclatural status, subject to the same rules and recommendations and available with original date and author for a taxon at any rank within the name-group (*q.v.*).

corr., correctus, -a, -um,: lat., corrected (by); used under Bot. to indicate a corrected orthography, the abbreviation corr. following the authority and preceding the name of the author who first effected the correction; indicates a 'justified emendation' in the sense of Zoo.; *see* emendation, justified.

corr., correxit: lat: he corrected.

correct, 20: (Bot., Bact.) of a name, that by which a taxon should properly be known; the equivalent of valid (*q.v.*) of Zoo.

cotype: a term, now superseded, formerly used for syntype, isotype or paratype (*q.v.*).

cryptogram: (Vir.) a shorthand summary of the properties of a virus or virus taxon, consisting of 4 pairs of symbols, describing, respectively: type of nucleic acid/strandedness of nucleic acid; molecular weight of nucleic acid (in millions)/ percentage of nucleic acid in infective particles; outline of particle/outline of nucleic acid and most closely surrounding protein (nucleocapsid); kind of host infected/kind of vector.

cultivar, 41: (Cult., Orch.) 1. A taxon consisting of an assemblage of cultivated plants, maintained in cultivation, and retaining its distinguishing features when reproduced.

   2. The corresponding taxonomic category. Abbreviated cv.

cultivars, names of, 42.

cultivar class, 43: (Cult., Orch.) the taxon within which the use of the same cultivar name for two different cultivars would cause confusion.

cultivarietas: lat., cultivar, *q.v.*

cultivated plants, names of, 41.

cv: cultivar, *q.v.*

date, publication, of, 35: 1. Of a work, the date of its becoming available to the general public or to relevant institutions.

   2. Of a name, the date on which the criteria of availability (Zoo.) or valid publication (Bact., Bot.) were first satisfied.

declaration, 7, 25: (Zoo.) a provisional amendment to the Code, issued by the International Commission on Zoological Nomenclature, and operative from its date of publication until the next succeeding International Zoological Congress (or authorized equivalent) ratifies, modifies or rejects it.

definition: (Zoo.) a statement of the features distinguishing a taxon; the equivalent of diagnosis (*q.v.*) of Bact. and Bot.

denomination, variety, 44; cultivar name.

descr., descriptio: lat., description, *q.v.*

description: a statement of the attributes of a specimen or taxon.

designation: 1. (Zoo.) the act of an author in fixing, by an express statement, the type of the name of a taxon of the species-group or the genus-group.

   2. (Bot.) a general term for that by which a taxon is referred to, e.g. a name or formula, *q.v.*

diagnosis: (Bot.) a statement of the features of a taxon which, in the opinion of the author, distinguish it from others; the equivalent of definition of (*q.v.*) Zoo.

direction: (Zoo.) a decision of the International Commission on Zoological Nomenclature, completing an earlier decision, or required as such by the provisions of
the Code.

double citation, 33.

doubtful name, 25; *see* nom. dub.

dubious name: *see* nom. dub.

ecosphere, 1: the planetary ecosystem, consisting of living organisms and the
components of the environment with which they react; the earth's life-support
system.

effective publication, 16: *see* publication, effective.

emend., emendatus, -a, -um: lat., altered (by); indicates a change in circumscription of a taxon without exclusion of the type of its name: the abbreviation emend.
follows the authority and precedes the name of the author who effected the
change.

emen.., emendavit: lat., he emended.

emendation: (Zoo.) 1. Any demonstrably intentional change in the spelling of a
name.

2. A name the spelling of which has undergone such a change.

emendation, justified: (Zoo.) one in accordance with the provisions of the Code.

emendation, unjustified: (Zoo.) one not so, to be treated as a new name with its own
authority and date.

entity, taxonomic: *see* taxon.

e.p., ex parte: lat., in part.

epithet, 34: (Bact., Bot.) a word, other than a generic name or a term indicative of
rank, forming part of a combination (*q.v.*).

epithet, collective: *see* collective.

epithet, specific: *see* specific epithet.

erect: *see* establish (2).

establish: 1. (Zoo.) Of a name, to make available.

2. Of a taxon, to describe and validly publish (or make available) a name for;
to erect.

ex, 32: lat., from, according to. 1. (Bot.) used to connect the names of two persons,
the second of which validly published a name proposed but not validly published
by the first.

2. (Bot.) used to connect the names of two persons, the second of which
effectively published, as a synonym, a name proposed in manuscript by the
first.

3. (Bact., Zoo.) used to connect the names of two persons, the first of which
published a name proposed but not published by the second.

exceptions, 49.

excl., exclusus, -a, -um: lat., excluded; used to indicate elements included in a
taxon by a previous author or authors, but considered not to belong to it by the
writer and excluded from it by him.

excl. gen., excluso genere: lat., with the genus excluded.

excl. spec., exclusa speciei: lat., with the species excluded.

excl. spec., excl. specim., exclusis speciminibus: lat., with the specimens excluded.

ex parte: lat., in part; *see* p.p.

f., fig., figura: lat., figure.

f., fil., filius: lat., son.

families, names of, 8.

family-group, 11: (Zoo.) the assembalge of categories from Tribe to Superfamily inclusive.

fixation (Zoo.): a general term for the determination of a type-species, whether by designation or indication (*q.v.*).

formae, names of, 10.

forma specialis: lat., special form; (Bact., Bot.) a variant of a parasitic or symbiotic species distinguished primarily by its adaptation or restriction to a particular host or hosts. Abbreviated f. sp.

formation: *see* formulation.

form-genus, 39: (Bot.) 1. A taxon of fossil plants of generic rank, which may be unassignable to a family, to which are referred apparently similar small isolated parts, e.g. leaf-fragments, roots, spores, seeds.

    2. A taxon of imperfect fungi of generic rank.

    The name of a form-genus can be used to refer only to the part or state represented by its type.

forms, names of, 11.

forms, names, of, 8.

formula, 45: (Bot., Cult., Orch.) a designation of a hybrid or hybrid group, formed by connecting the names of the parents by a sign (× or +) or by combining them into one, forming a condensed formula preceded by a similar sign, e.g. *Crataegus* × *Mespilus*; × *Crataegomespilus*. *See also* formula, condensed.

formula, condensed, 45: a formula formed by combining parts of the names of two genera, and applied to intergeneric hybrids between them.

formulation, 15: the way in which a name is formed.

fossil: an organism, or part of it, preserved by some natural means in the geological record, or an impression or petrifaction so preserved; *contrast* recent (*q.v.*).

fossil plants, names of, 40.

f. sp.: *see* forma specialis.

fungi, names of, 40.

fungi, imperfect, names of, 40.

genera, names of, 8.

generitype: the type of the name of a genus.

genotype: 1. The hereditary or genetic constitution of an individual.

    2. *See* generitype.

genus-group, 11: (Zoo.) the categories genus and sub-genus.

genus, name of, 8.

graft chimaeras, names of: *see* graft hybrids.

graft hybrids, names of, 45.

grex, 46: (Cult., Orch.) a hybrid taxon to which are referred all the progeny arising from any, each and every crossing of any two parent plants belonging to different taxa that bear the same pair of specific, hybrid-specific or grex names.

group: 1. (Bact.) An informal taxon, based upon antigenic analysis, consisting of an assemblage of serotypes (*q.v.*).

    2. (Zoo.) An assemblage of nomenclaturally coordinate categories; a name-group.

3. (Cult.) An assemblage of similar cultivars within a species or interspecific hybrid.

4. (Vir.) A viral taxon of uncertain rank, as yet not designated either as a family or as a genus.

group, collective: *see* collective group.

group, taxonomic: *see* taxon.

Handbook on Orchid Nomenclature and Registration, 46.

heterotypic: *see* synonym, heterotypic.

hierarchal: *see* hierarchical system.

hierarchical system, 2.

hierarchy, taxonomic, 4: the framework formed by the conventional ordering of taxonomic categories into a series of consecutively subordinate levels or ranks.

holotype, 20: the sole element used as the type (*q.v.*) by the author of a name or the one element designated or indicated by him as the type or holotype.

homonym, 23: a name identical in orthography with another (or treated as such by the appropriate Code) and based on a different type.

homonym, junior: (Zoo.) the later published of two homonyms.

homonym, primary: (Zoo.) any of two or more species-group homonyms applied to taxa referred to the same nominal genus when first published.

homonym, secondary: (Zoo.) any of two or more species-group homonyms referred to the same nominal genus as a result of the transfer of one or more taxa from another genus or other genera.

homonym, senior: (Zoo.) the earlier published of two homonyms.

homonymy: the existence of two or more names of identical orthography based on different types.

homotypic: *see* synonym, homotypic.

hort., hortulanorum: (Bot.,) lat., of gardeners; used in citation to denote a name of garden origin.

hybrids, animal, names of, 44.

hybrids, orchid, names of, 45, 46.

hybrids, plant, graft, names of, 45.

hybrids, plant, sexual, names of, 45.

ib., ibidem: lat., the same, in the same place.

illegitimacy, 23: 1. (Bot., Bact.) a state of non-accordance with the rules that requires a name not to be taken into consideration for the purposes of priority (except for the purposes of homonymy, *q.v.*) when the correct name of a taxon is being decided.

2. (Bact., Bot., Cult., Vir.) a state of non-accordance with the articles of the Code.

illegitimate name, 23: *see* nom. illegit.

imperfect fungi, names of, 40.

in, 32: lat., in; used to connect the names of two persons, the second of which was the editor, or overall author, of a work in which the first was responsible for validly publishing, or making available, a name.

inadmissible: (Bot.) of a name, of a form that precludes its valid publication (*q.v.*); of the use of a name or epithet, contrary to the provisions of the Code.

incertae cedis: lat., of uncertain seat, i.e. of uncertain taxonomic position.

index, official, 25: (Zoo.) an index of names considered, rejected and invalidated by
the International Commission on Zoological Nomenclature, published by the
International Trust for Zoological Nomenclature; indexes exist for specific
generic and family-group names and also for rejected and invalid works.

indication: (Zoo.) 1. Published information serving in lieu of a description or
definition (*q.v.*) as a criterion of availability (before 1931).

2. Published information determining the type-species of a generic name in the
absence of an original designation (*q.v.*).

ined., ineditus, -a, -um, : lat., unpublished.

International Code of Botanical Nomenclature, 5.

International Code of Botanical Nomenclature, modification of, 6.

International Code of Nomenclature of Bacteria, 5.

International Code of Nomenclature of Bacteria, modification of, 6.

International Code of Nomenclature of Cultivated Plants, 42.

International Code of Nomenclature of Cultivated Plants, modification of, 44.

International Code of Zoological Nomenclature, 5.

International Code of Zoological Nomenclature, modification of, 6.

interspecific: (Bot.) of a hybrid, between species referred to the same genus.

invalid: 1. (Bact.) Of names, not validly published; the equivalent of not validly
published of Bot. and of unavailable of Zoo.

2. (Zoo.) Of names, available (*q.v.*) for a given taxon but not that by which it
should properly be known; the equivalent of incorrect of Bact. and Bot.

3. Of taxa, not recognized as taxonomically distinct, at least at the rank in
question.

isosyntype: a duplicate of a syntype, not cited in the protologue (*q.v.*).

junior homonym: *see* homonym, junior.

junior synonym: *see* synonym, junior.

kingdom, 3: the highest generally employed category of the taxonomic hierarchy.
The Code of Nomenclature to be applied to the naming of any taxon of organisms is determined by which of the following four major groups—animals,
plants, bacteria and viruses—it is considered to belong to. For taxonomic
purposes a varying number of kingdoms may be recognized, e.g. 5—R. H.
WHITTAKER in *Science, N. Y.*, **163**: 150–63 (1969) or 7—C. JEFFREY in *Kew Bull.*,
25: 296–9 (1971).

language, names, of, 7.

l.c., loc. cit., loco citato: lat., in the place cited; used to avoid the repetition of a
bibliographic reference already given.

lectotype, 20: an element subsequently designated or selected from amongst syntypes to serve as the definitive type.

legitimacy, 23: 1. (Bact., Bot.) the state of accordance with the rules that requires a
name to be taken into consideration for the purposes of priority when the correct
name of a taxon is being decided.

2. (Bact., Bot., Cult., Vir.) the state of accordance with the articles of the
Code.

legitimate name: *see* nom. legit.

nec: lat., and not (of); nor (of).

neotype, 20: an element designated or selected subsequently to serve as the type of a name when all the original·type elements are destroyed or missing or believed so to be.

new name: *see* nom. nov.

nom. ambig., nomen ambiguum, 25: lat., ambiguous name; a name used in different senses—i.e. applied by different authors to different taxa—so that it has become a long-persistent source of error.

nom. conf., nomen confusum, 25: lat., confused name; a name based on a type consisting of discordant elements from which it is impossible to select a satisfactory lectotype; in Bact., specifically a name the type of which was an impure or mixed culture.

nom. cons., nom. conserv., nomen conservandum, 25: lat., a name to be conserved; a conserved name, i.e. a name the use of which has been officially sanctioned in spite of its contravention of one or more provisions of a Code; the procedure of giving sanction is known as conservation (*q.v.*).

nom. dub., nomen dubium, 25: lat., dubious name; a name of uncertain application, either because, through lack of the original type and sufficient information about it, satisfactory typification is impossible, or because it is impossible to ascertain to which taxon its type should be referred.

nom. illegit., nomen illegitimum: lat., illegitimate name: 1. (Bact., Bot.) a validly published name that is not in accordance with the rules in such a way that it must not be taken into consideration for the purposes of priority (except for the purposes of homonymy, *q.v.*) when the correct name of a taxon is being decided.

2. (Bact., Bot., Cult., Vir.) a name that is not in accordance with the articles of the Code.

nom. legit., nomen legitimum: lat., legitimate name: 1. (Bact., Bot.) a validly published name that is in accordance with the rules in such a way that it must be taken into consideration for the purposes of priority when the correct name of a taxon is being decided.

2. (Bact., Bot., Cult., Vir.) a name that is in accordance with the articles of the Code.

nom. non rite public., nomen non rite publicatum: lat., name not properly published; used in citation to indicate a name that has been effectively but not validly published.

nom. nov., nomen novum: lat., new name; a name expressly proposed and published to replace an earlier name that cannot be used for some reason, e.g. if it is a later (junior) homonym.

nom. nud., nomen nudum: lat., naked name; a name published without such associate descriptive matter as is required by the appropriate Code to satisfy the criteria of availability (Zoo.) or valid publication (Bact., Bot.).

nom. oblit.: *see* nomen oblitum.

nom. rejic., nomen rejiciendum, 25: lat., a name to be rejected; a rejected name, i.e. a name the use of which has been officially rejected, usually in favour of another (conserved) name; under Bact. and Zoo. names listed as officially rejected are to be permanently rejected; under Bot., rejected earlier homonyms and nomenclatural synonyms are to be permanently rejected, as are names rejected as widely and persistently misapplied (*see* §5.16, p. 24), but rejected taxonomic synonyms are to be rejected only for so long as they are considered to apply to the same taxon as the corresponding conserved name; otherwise the normal rules of priority apply.

nom. superfl., nomen superfluum, 24, 52: lat., superfluous name; (Bot., Bact.) a name which when first validly published, was applied by its author to a taxon so circumscribed as to include the type of another name which the author ought to have adopted under the rules.

nomen: lat., a name.

nomen approbatum: lat., approved name. (Bact.) *see* approved.

nomenclatural synonym: *see* synonym, homotypic.

nomenclature, 2: 1. The giving of names to taxa.

2. The system of names so produced, or any part thereof; the result of 1.

nomenclature, alphabet of, 8.

nomenclature, classification, and, independence of, 2.

nomenclature, classification, dependence of on, 30.

nomenclature, codes of, 5.

nomenclature, instability of, 13.

nomenclature, language of, 8.

nomen hybridum: lat., hybrid name. (Bact.) a name formed by combining words derived from different languages.

nomen oblitum: lat., forgotten name. (Zoo.) a term formerly employed in Zoo. (article 23b) to denote a name that had remained unused as a senior synonym in the primary zoological literature for more than 50 years. Article 23b was replaced by Declaration 43 of the International Commission on Zoological Nomenclature —(1970) *Bull. Zool. Nom.*, **27**: 135—whereby its interpretation during its period of operation (1 January 1961 to 3 December 1970 inclusive) was also clarified. This repeal was accepted by the 17th International Congress of Zoology at Monaco (1972) but the Declaration was modified and the International Code of Zoological Nomenclature amended such that Articles 23(a) and 23(b) of the 1964 Code were replaced by a combined Article 23(a–b)—see *Bull. Zool. Nom.*, **31**: 79–80 (1974). *Nomina oblita* are thus no longer recognized by the Code and a *nomen oblitum* is now defined as an unused senior synonym placed on an official index of rejected names under the provisions of the former Article 23b during the period of its operation.

nomen perplexum, 25: lat., perplexing name. (Bact.) a name the application of which is known but which causes uncertainty in bacteriology.

nomen revictum, 52: lat., revived name. (Bact.) a name validly published before 1 Jan. 1980 but not included in an Approved List and revived by an author after 1 Jan. 1980 as a new name and applied by the later author to a taxon with the same circumscription, position and rank as given by the original author.

nomen triviale: lat., trivial name; the specific name (Zoo.) or specific epithet (Bact., Bot.), *q.v.*

nominal: (Zoo.) of a taxon bearing a given name, in the sense of the type.

nominate, 39: (Zoo.) containing the type of the name of the higher taxon to which it is subordinate.

nominifer: a type: *see* type 1).

nm.: *see* nothomorph.

non: lat., not (of).

non . . . nec: lat., neither (of) . . . nor (of).

non-statutory: (Cult.) established by voluntary agreement between organizations.

nothomorph: (Bot.) 1. A taxonomic category subordinate to the collective (hybrid) category equivalent to species and the equivalent of variety in the taxonomic hierarchy.

  2. Any hybrid variant derived from the same parent species, forming a taxon of this category. Abbreviated nm.

objective synonym: *see* synonym, homotypic.

obligate synonym: *see* synonym, homotypic.

oldest: first validly published (Bact., Bot.) or first made available (Zoo.).

op. cit., opero citato: lat., in the work cited; used to avoid the repetition of part of a bibliographic reference already given.

opinion, 25: (Bact., Zoo.) a decision of the International Commission on Zoological Nomenclature, or of the Judical Commission of the International Committee on Systematic Bacteriology, on any particular case referred to it.

orchids, cultivated, names of, 46.

organ-genus: a category formerly employed in Bot. for genera of fossil plants, the characters of which were derived principally from a single organ, e.g. fructification, stem, and which were assignable to a family; it is now included in the category form-genus (q.v.).

orthographic variants, 26: different spellings of the same name.

orthography: spelling.

orthography, names, of, 26.

orth. mut., orthographia mutata: lat., with an altered spelling (by).

paralectotype: (Zoo.) a remaining syntype after a lectotype has been designated from amongst syntypes; not a type in the strict nomenclatural sense.

paranym: a name so similar in orthography to another (based on a different type) that the two are likely to be confused.

paratype: 1. A specimen cited in the protologue (q.v.) other than the holotype and isotype(s) or syntypes.

2. (Bot.) Also, a remaining syntype after a lectotype has been chosen from amongst syntypes.

A paratype is not a type in the strict nomenclatural sense.

parentheses, use of in citation, 33.

phagotype: (Bact.) a variant of a species distinguished by its sensitivity to a particular bacteriophage or by a distinctive pattern of sensitivity to a set of specific bacteriophages. The use of the term 'phagovar' to replace 'phagotype' is recommended by the ICNB.

phase: (Bact.) in *Enterobacteriaceae*, a well-defined stage of a naturally-occurring alternating variation.

plants, names of, 5.

plants, cultivated, names of, 41.

plants, fossil, names of, 40.

position, 12: the place of a taxon in a system of classification, indicated by the higher taxon to which it is referred.

p.p., pro parte: lat., in part; used in citations to show that only a part of a taxon as circumscribed by a previous author is being referred to by the writer.

preoccupied: (Zoo.) already in use for another taxon with a different type.

primary homonym: *see* homonym, primary.

priority, 20: precedence in date of availability (Zoo.) or valid publication (Bact., Bot., Cult.).

priority, limitations of, 21.

pro hybr., pro hybrida: lat., as a hybrid; used to indicate that a name of a taxon regarded as a species was originally published as a name of a hybrid.

pro parte: *see* p.p.

pro spec., pro speciei: lat., as a species; used to indicated that a name of a taxon regarded as a hybrid was originally published as a name of a species.

pro syn., pro synonymo: lat., as a synonym; used to indicate that a name was originally published in synonymy, as a synonym.

protologue: the whole of the verbal and illustrative matter associated with a name at its place of first valid publication.

publication, 16: 1. The process of distributing graphic material in a way that satisfies the criteria of publication of the appropriate Code.

2. Any item so published under 1.

3. in Zoo. also a condition to be satisfied in order that a name may become available; the equivalent of effective publication of Bot. and Bact.; *see* publication, effective.

publication, effective, 16: (Bact., Bot.) the process of distributing graphic material of a kind and in a way that satisfies the criteria of effective publication of the appropriate Code.

publication, effective, criteria of, 16.

publication, valid, 18: (Bact., Bot., Cult.) the publication of names in accordance with the criteria of valid publication of the appropriate Code; names that have not undergone valid publication are treated as non-existent for the purposes of nomenclature.

publication, valid, criteria of, 17.

quoad: lat., as to (the), as regards (the); with respect to (the); used in citation to indicate what part of a taxon as circumscribed by a previous author is being referred to by the writer.

rank, 4: the level in the taxonomic hierarchy of a category and of the taxa of that category.

rank, taxonomic, 4: *see* rank.

recent, 21: not fossil; living.

recommendation, 5: a non-mandatory provision of a Code, showing what is considered to be a good practice, but which can be ignored without contravention of the Code.

registration, 44: 1. (Cult.) The placing by a nationally or internationally recognized body (the registration authority) of a cultivar name on to an official list or register.

2. (Orch.) The likewise placing of a grex name.

rejected: *see* rejected name.

rejected name: 1. *See* nom. rejic.

2. (Bot.) Any name applied to a given taxon other than the correct one, e.g. names that are inadmissible, not validly published, are illegitimate for some reason, are rejected names in the sense of 1, or are synonyms of the correct name.

3. (Zoo.) Any name not employed by an author for any reason.

replacement name: (Zoo.) 1. A nomen novum (*see* nom. nov.).

2. An available synonym adopted to replace an earlier preoccupied name (*q.v.*).

rule, 5: a mandatory provision of a code of nomenclature.

rite: lat., properly, correctly, according to the rules.

saltem: lat., at least.

sched., scheda: lat., label (of a specimen).

sec., secundum: lat., according to.

secondary homonym: *see* homonym, secondary.

senior homonym: *see* homonym, senior.

senior synonym: *see* synonym, senior.

sens. str.: *see* s.s.

sens. lat.: *see* s.l.

sensu amplo: *see* s.l.

serotype: (Bact.) a variant of a species or subspecies distinguished from other such variants of the same species or subspecies on the basis of its antigenic structure. The use of the term 'serovar' to replace 'serotype' is recommended by the ICNB.

sigla (Vir.): names made up from a few, generally initial, letters (e.g. *Reovirus* from *r*espiratory *e*nteric *o*rphan viruses); such names may be employed as the names of viruses or viral taxa, provided that they are meaningful to workers in the fields and are recommended by international virus study groups.

sine: lat., without.

s.l., sens. lat., sensu lato: lat., in the broad sense; i.e. of a taxon, including all its subordinate taxa and/or other taxa sometimes considered as distinct.

species, 2.

species, names of, 10.

species-group, 11 (Zoo.): the categories species and subspecies.

species inquirenda: (Zoo.) lat., a species to be queried; a doubtfully identified species requiring further investigation.

specific epithet, 51: (Bact., Bot.) the second word in the name of a species; the equivalent of specific name of Zoo.

specific name, 51: 1. (Zoo.) The second word in the name of a species; the equivalent of specific epithet of Bot. and Bact.

2. The name of a species.

sphalm., sphalmate: lat., by mistake, in error.

s.s., sens. str., sensu stricto: lat., in the strict sense, in the narrow sense, i.e. of a taxon, in the sense of the type of its name; or in the sense of its circumscription by its original describer; or in the sense of its nominate subordinate taxon (in the case of a taxon with 2 or more subordinate taxa); or with the exclusion of similar taxa sometimes united with it.

stage: (Bact., Bot.) *see* state.

starting-point, 21: a published work, and the date thereof, before which no name is considered to have been made available (Zoo.) or validly published (Bact., Bot., Cult.).

state: 1. (Bact.) One of certain variants which arise in cultures of many bacteria and which bring about a change in the gross appearance of a culture.

2. (Bot.) A phase in a life-cycle.

stat. nov., status novus: lat., new status; 1. used in citation to indicate that a taxon has been altered in rank but retains in its name the epithet from its name in the former rank.

2. Used in citation to indicate that a taxon has been changed in status, from specific to hybrid or vice versa.

status: 1. the nomenclatural standing of a name: a name that has not satisfied the criteria of availability (Zoo.) or valid publication (Bact., Bot.) has no nomenclatural standing, and is treated as non-existent for the purposes of nomenclature.

2. The property of a name that corresponds to rank as a property of a taxon.

3. The condition of a taxon as of hybrid or non-hybrid nature.

statutory: (Cult.) established by a legal enactment or process of a country, or by legal treaty between countries.

strain: (Bact.) the descendants of a single isolation in pure culture, sometimes showing marked differences in economic significance from other strains or isolations; analogous to clone (*q.v.*) of Bot. and Cult.

subfossil: (Bot.) fossil (*q.v.*) but geologically young, soft in texture, organic in composition, and usually found in a soft deposit, such as peat; the equivalent of fossil for nomenclature purposes.

subjective synonym: *see* synonym, heterotypic.

subspecies, names of, 11.

subspecific epithet, 51: (Bact., Bot.) the third word (other than a word indicative of rank) in the name of a subspecies; the equivalent of subspecific name of Zoo.

subspecific name, 51; (Zoo.) the third word in the name of a subspecies; the equivalent of subspecific epithet of Bact. and Bot.

substitute name: *see* replacement name.

superfluous, 24: (Bact., Bot.) applied by its author to a taxon so circumscribed by him as to include the type of another name which could and should have been used as its correct name; *see also* nom. superfl.

superfluous names, 24: *see* superfluous.

suppress: (Zoo.) of a name, to make effectively unavailable (reject), for the purpose prescribed; only the International Commission on Zoological Nomenclature, by use of its Plenary Powers, has authority to suppress a name. *See* nom. rejic., suppression.

suppressed: (Zoo.) made effectively unavailable (rejected) for the purposes prescribed, by action of the International Commission on Zoological Nomenclature under its Plenary Powers. *See* nom. rejic.

suppression: (Zoo.). An act of the International Commission on Zoological Nomenclature, using its Plenary Powers, by which a name is made effectively unavailable (rejected) for such purposes as the Commission may prescribe (homonymy and/or priority). *See* nom. rejic.

syn., synonymum: lat., synonym. 1. *See* synonym.

　　2. In the abbreviated form, used before a name to indicate it is a synonym of the name properly to be used for the taxon concerned.

synonym, 28: 1. One of two or more names applied to the same taxon.

　　2. Commonly, a name applied to a given taxon other than the one by which it should properly be known; a name placed in, and forming part of, the synonymy thereof.

synonym, absolute: *see* synonym, homotypic.

synonym, commercial: (Cult.) a name that may properly be used for a cultivar in place of the internationally correct one in countries where the latter is commercially unacceptable.

synonym, heterotypic, 29: a synonym based on a different type (from that of another).

synonym, homotypic, 29: a synonym based on the same type as another.

synonym, junior, 28: (Zoo.) the later published of two synonyms.

synonym, nomenclatural: *see* synonym, homotypic.

synonym, objective: *see* synonym, homotypic.

synonym, obligate: *see* synonym, homotypic.

synonym, senior, 28: (Zoo.) the earlier published of two synonyms.

synonym, subjective: *see* synonym, heterotypic.

synonym, taxonomic: *see* synonym, heterotypic.

synonymy, 28: 1. The existence of two or more names applied to the same taxon.

　　2. The relationship between any two such names.

　　3. The names considered to apply to a given taxon other than the name by which it should properly be known.

　　4. A list of such names.

syntype, 20: any of two or more elements used as types by the author of a name, whether or not designated as such by him.

system: the result of a given classification of a number of living organisms, usually one dealing with a taxon of high rank, e.g. Kingdom, Phylum, Division: a system of classification.

system, hierarchical, 2.

system, taxonomic: *see* system.

systematics, 1: the scientific study of the variation of living organisms.

t., tab., tabula: lat., plate.

tantum: lat., only, merely.

tautonym, 24: a name of a species in which the second word exactly repeats the generic name.

tautonymy: the existence of a tautonym; *see* tautonym.

taxa: plural of taxon, *q.v.*

taxa, definition of, 13.

taxa, names of, 8.

taxon, 2: a taxonomic group (or unit, or entity) of any rank; in plural, taxa.

taxon, subordinate, nominate, 37: *see* nominate.

taxonomic category, 4: *see* category, taxonomic.

taxonomic entity, 50: *see* taxon.

taxonomic group, 50: *see* taxon.

taxonomic hierarchy, 4: *see* hierarchy, taxonomic.

taxonomic position: *see* position.

taxonomic rank, 4: *see* rank.

taxonomic synonym: *see* synonym, heterotypic.

taxonomic unit, 50: *see* taxon.

taxonomy, 50: 1. The study of the principles and practice of classification.

   2. Loosely, systematics, *q.v.*

ternary: *see* trinominal.

tom. cit., tomo citato: lat., in the volume cited; used to avoid the repetition of part of a bibliographic reference already given.

topotype: a specimen from the same locality as a type; not a type in the strict nomenclatural sense.

transfer: the alteration of the position of a taxon.

trans. nov., translatio nova: lat., new transfer; used in citation to indicate that a taxon has been altered in position but retains in its name the epithet from its name in the former position.

trinomen, 51: (Zoo.) the name of a subspecies, consisting of the name of the species in which it is classified, followed by a word peculiar to the subspecies; the equivalent of subspecific name of Bact. and Bot.

trinomial: 1. *See* trinominal.

   2. A name consisting of three terms, any of which can consist of any number of words; when each term consists of only one word, then a name such as the name of a subspecies (Bact., Bot., Zoo.) or infrasubspecific taxon (Bot. only). *See* trinominal.

trinominal, 11: 1. Consisting of three words, such as the name of a subspecies.

   2. Making use of names consisting of three words.

trivial name: a specific epithet (Bact., Bot.) or specific name (Zoo.); lat., nomen triviale.

typ. cons., typus conservandus: lat., a type to be conserved.

type, 18: 1. (Bact., Bot., Zoo.) An element on which the descriptive matter fulfilling the conditions of availability (Zoo.) or valid publication (Bact., Bot.) for a name is based, or is considered to have been based, and which by its taxonomic position decides the application of the name.

   2. (Bact.) A variant within a species, showing certain distinct features, but of

lower than subspecific rank; the use of the term in this sense is not recommended by the ICNB, even in compounds such as morphotype, serotype, *q.v.*, in which it has long been used; the use of the suffix -var is recommended instead, as in e.g. morphovar, serovar.

type, nomenclatural: *see* type 1).

type method, 18: the method of deciding the application of names by means of types; *see* type 1).

typification, 18: the designation of a type; *see* type 1).

typify: 1. To designate the type of a name.

    2. To be the type of a name.

typotype: the type of a type; e.g. if the type of name, studied by an author, is a description or illustration previously published by an earlier author, then the element on which the earlier author's description or illustration was based, and which, as such, the later author did not study, is the typotype of the later author's name.

uninomial: 1. *See* uninominal.

    2. A name consisting of one term, which can be of any number of words; when the term consists of only one word, then a name such as the name of a genus, family or kingdom.

uninominal, 8: 1. Consisting of one word, such as the name of a genus, family or kingdom.

    2. Making use of names consisting of one word.

unit, taxonomic: *see* taxon.

unitary: *see* uninominal.

v., vel: lat., or.

valid, 21: 1. (Zoo.) Of a name, that by which a taxon should properly be known; the equivalent of correct (*q.v.*) of Bot. and Bact.

    2. (Bact.) Validly published.

    3. Of a taxon, 'good' in the sense of generally accepted as distinct by competent authorities.

validated: (Zoo.) considered and accorded valid status by the International Commission on Zoological Nomenclature.

validated names, 25: *see also* validated.

valid publication, 18: *see* publication, valid.

validly published: *see* publication, valid.

variants, orthographic, 26: different spellings of the same name.

varieties, names of, 11 (Bot.); 42 (Cult.): *See also* variety.

variety, 11, 42: 1. (Bot.) The category varietas.

    2. (Bot.) A taxon of this category.

    3. (Bact.) Another name for the category subspecies.

    4. (Bact.) A taxon of this category.

    5. (Cult.) The category cultivar.

    6. (Cult.) A taxon of this category.

vernacular names, 5, 50.

verosim., verosimiliter: lat., probably.

viruses, names of, 47.

work: (Zoo.) of an animal, a result of its activities, but not a part of it; e.g. worm-tube, track.